Sloshing in Upright Circular Containers

Mathematics and its Applications

Modelling, Engineering, and Social Sciences

Series Editor: Hemen Dutta, Department of Mathematics, Gauhati University

Tensor Calculus and Applications
Simplified Tools and Techniques
Bhaben Kalita

Discrete Mathematical Structures
A Succinct Foundation
Beri Venkatachalapathy Senthil Kumar and Hemen Dutta

Methods of Mathematical Modelling
Fractional Differential Equations
Edited by Harendra Singh, Devendra Kumar, and Dumitru Baleanu

Mathematical Methods in Engineering and Applied Sciences
Edited by Hemen Dutta

Sequence Spaces
Topics in Modern Summability Theory
Mohammad Mursaleen and Feyzi Başar

Fractional Calculus in Medical and Health Science
Devendra Kumar and Jagdev Singh

Topics in Contemporary Mathematical Analysis and Applications
Hemen Dutta

Sloshing in Upright Circular Containers
Theory, Analytical Solutions, and Applications
Ihor Raynovskyy and Alexander Timokha

For more information about this series, please visit:
www.crcpress.com/Mathematics-and-its-applications/book-series/MES

Sloshing in Upright Circular Containers
Theory, Analytical Solutions, and Applications

Ihor Raynovskyy
Alexander Timokha

CRC Press
Taylor & Francis Group
Boca Raton London New York

CRC Press is an imprint of the
Taylor & Francis Group, an **informa** business

First edition published 2021
by CRC Press
6000 Broken Sound Parkway NW, Suite 300, Boca Raton, FL 33487-2742

and by CRC Press
2 Park Square, Milton Park, Abingdon, Oxon, OX14 4RN

© 2021 Taylor & Francis Group, LLC

CRC Press is an imprint of Taylor & Francis Group, LLC

Reasonable efforts have been made to publish reliable data and information, but the author and publisher cannot assume responsibility for the validity of all materials or the consequences of their use. The authors and publishers have attempted to trace the copyright holders of all material reproduced in this publication and apologize to copyright holders if permission to publish in this form has not been obtained. If any copyright material has not been acknowledged please write and let us know so we may rectify in any future reprint.

Except as permitted under U.S. Copyright Law, no part of this book may be reprinted, reproduced, transmitted, or utilized in any form by any electronic, mechanical, or other means, now known or hereafter invented, including photocopying, microfilming, and recording, or in any information storage or retrieval system, without written permission from the publishers.

For permission to photocopy or use material electronically from this work, access www.copyright.com or contact the Copyright Clearance Center, Inc. (CCC), 222 Rosewood Drive, Danvers, MA 01923, 978-750-8400. For works that are not available on CCC please contact mpkbookspermissions@tandf.co.uk

Trademark notice: Product or corporate names may be trademarks or registered trademarks, and are used only for identification and explanation without intent to infringe.

Library of Congress Cataloging-in-Publication Data
Names: Raynovskyy, Ihor, author. | Timokha, A. N., author.
Title: Sloshing in upright circular containers : theory, analytical
  solutions and applications / Ihor Raynovskyy and Alexander Timokha.
Description: First edition. | Boca Raton : CRC Press, 2020. |
Series: Mathematics and its applications: modelling, engineering, and social
  sciences, 2689-0232 | Includes bibliographical references and index.
Identifiers: LCCN 2020022752 (print) | LCCN 2020022753 (ebook) | ISBN
  9780367362898 (hbk) | ISBN 9780429356711 (ebk)
Subjects: LCSH: Sloshing (Hydrodynamics) | Storage tanks.
Classification: LCC TA357.5.S57 R39 2020 (print) | LCC TA357.5.S57
  (ebook) | DDC 620.1/064--dc23
LC record available at https://lccn.loc.gov/2020022752
LC ebook record available at https://lccn.loc.gov/2020022753

ISBN: 978-0-367-36289-8 (hbk)
ISBN: 978-0-429-35671-1 (ebk)

Typeset in CMR
by Nova Techset Private Limited, Bengaluru & Chennai, India

Contents

Preface ix

Authors xi

Chapter 1 Introduction 1

    1.1 Origins of the multimodal method 2
        1.1.1 Linear sloshing 2
        1.1.2 Nonlinear sloshing 4
        1.1.3 Perko's method 5
    1.2 The nonlinear modal modelling 5
    1.3 Nonlinear modal systems and steady-state analysis 7
        1.3.1 Upright tanks 7
            1.3.1.1 Two-dimensional sloshing in rectangular tanks 7
            1.3.1.2 Three-dimensional sloshing in rectangular tanks 9
            1.3.1.3 Annular and circular tanks 9
        1.3.2 Nonconformal mapping technique 10
    1.4 Challenges of the multimodal modelling 11

Chapter 2 Differential, variational, and modal statements 13

    2.1 Free-surface problem 13
        2.1.1 Governing equations 14
        2.1.2 Boundary conditions 15
        2.1.3 Mass conservation 16
        2.1.4 Initial and periodicity conditions 17
    2.2 Bateman-Luke variational principle 17
        2.2.1 Irrotational flows 17
        2.2.2 Rotational flows 19
    2.3 Miles-Lukovsky modal equations 23
        2.3.1 Modal representation of the solution 23
        2.3.2 Euler-Lagrange equation 24
        2.3.3 Lukovsky's formulas for the hydrodynamic force and moment 28

Chapter 3 Linear modal theory and damping 33

    3.1 Linear natural sloshing modes and frequencies 33
        3.1.1 Theory 33
        3.1.2 Exact solutions and associated wave types 35
            3.1.2.1 Rectangular tank 35
            3.1.2.2 Upright circular tank 36
            3.1.2.3 Annular and sectored upright circular tank 38
    3.2 Linear sloshing problem and its modal solution 39
        3.2.1 Linear multimodal method 40
            3.2.1.1 Linearised sloshing problem 40
            3.2.1.2 Linear modal equations 40
            3.2.1.3 Steady-state and transient waves 42
            3.2.1.4 Lukovsky's formulas for the hydrodynamic force and moment 44
        3.2.2 Linear multimodal theory for an upright circular tank 46
            3.2.2.1 Hydrodynamic coefficients 46
            3.2.2.2 Modal equations 47
            3.2.2.3 Hydrodynamic force and moment 48
    3.3 Linear viscous damping 48
        3.3.1 Damped harmonic oscillator 48
        3.3.2 Viscous laminar boundary-layer and bulk damping 50
    3.4 Damping versus surface tension 52

Chapter 4 Nonlinear modal theories of nonparametric resonant waves in an upright circular container 55

    4.1 Nonlinear modal equations for small-amplitude three-dimensional tank motions 55
        4.1.1 Introductory remarks 55
        4.1.2 Natural sloshing modes and their normalisation 57
        4.1.3 Miles-Lukovsky modal system 58
    4.2 Adaptive modal equations (a weakly-nonlinear modal theory of the third order) 60
    4.3 Narimanov-Moiseev-type modal theory 70
        4.3.1 Narimanov-Moiseev-type modal equations 70
        4.3.2 Comparison with existing modal equations 73
        4.3.3 Secondary resonance 74
        4.3.4 Damping 76
    4.4 Hydrodynamic force and moment 76
        4.4.1 Force 76

		4.4.2	Moment 77
			4.4.2.1 Approximate generalised Stokes-Joukowski potentials 78
			4.4.2.2 Adaptive approximation of the hydrodynamic moment 83
			4.4.2.3 Narimanov-Moiseev approximation ... 85

Chapter 5 Narimanov-Moiseev–type equations and resonant steady-state waves 87

    5.1 Asymptotically equivalent orbital tank motions 87
    5.2 Periodic solutions of the Narimanov-Moiseev–type modal equations 89
        5.2.1 The lowest-order terms; secular equations 89
        5.2.2 Stability 91
        5.2.3 Summarising theorem 92
        5.2.4 The third-order components 92
    5.3 Alternative secular equations 95
        5.3.1 Undamped sloshing 95
        5.3.2 Damped sloshing 96

Chapter 6 Steady-state resonant sloshing due to the longitudinal forcing 99

    6.1 Steady-state wave modes 99
        6.1.1 Undamped sloshing 99
        6.1.2 Damped sloshing 100
    6.2 Response curves 102
        6.2.1 Wave amplitudes 103
        6.2.2 Phase lag 104
        6.2.3 Effective frequency domains of steady-state wave modes 105
    6.3 Comparison with experiments 107
        6.3.1 Phase lags 107
        6.3.2 Wave elevations near the wall and hydrodynamic force 109

Chapter 7 Steady-state sloshing for the orbital forcing 111

    7.1 Wave-amplitude parameters 111
    7.2 Wave-amplitude response curves for elliptic orbits 114
        7.2.1 Undamped sloshing 114
        7.2.2 Damped sloshing 117
    7.3 Phase lags for the damped sloshing 117
    7.4 Rotary tank motions 119

Chapter 8 Tables of hydrodynamic coefficients 123

References 141

Index 155

Preface

Owing to great interest in spacecraft technologies (fuel and oxygen tanks, etc.), designing storage and operational containers for oil, water, and liquefied natural gas (LNG), studies on resonant sloshing in a vertical cylindrical tank have been a challenge from the 1960s through 1980s. Over the past ten years, they also became relevant to bio- and pharmaceutical technologies (e.g., production of protein in orbitally shaken bioreactors), as well as an important element of fish cage engineering.

Mathematically, sloshing deals with a special class of the free-surface boundary value problem whose formulation most frequently assumes an ideal incompressible liquid with irrotational flows. Accounting for the viscous damping and other physical peculiarities of the contained liquid may, however, matter, especially for biotechnological applications. Containers with a bio-liquid 'wing'-along complex and closed three-dimensional trajectories, cyclical over a long time so that even being relatively small, the viscous damping eliminates the transient-wave component and, as a consequence, sloshing reaches a steady-state (time-periodic) wave mode, if, of course, this wave mode is stable. The steady-state sloshing problem normally does not have a unique solution and, therefore, the stable wave modes can co-exist for the same input physical parameters. Using computational fluid dynamics (CFD) simulations with different initial scenarios (solving the corresponding Cauchy problems) is not the optimal way to detect all the co-existing stable steady-state waves. An alternative could be analytical and semi-analytical approaches utilising a variational formalism for reduction of the original free-surface problem to a system of ordinary differential (modal) equations (the Euler-Lagrange–type equation) with respect to the hydrodynamic generalised coordinates and velocities. Dealing with this system instead of the original free-surface boundary value problem significantly simplifies the steady-state wave analysis. This is true even though the hydromechanical system has an infinite number of degrees of freedom, and therefore, the system of ordinary differential equations is infinite-dimensional; it needs a further simplification to admit analytical studies. Such a simplification can be based on introducing the special asymptotic relationships between the hydrodynamic generalised coordinates and velocities. The relationships help to exclude the generalised coordinates, which give a negligible contribution as well as to rewrite the modal system in a rather compact weakly nonlinear and often finite-dimensional form.

The above-described analytical scheme (variational principle $\to$ Euler-Lagrange–type [modal] equation $\to$ simplified weakly nonlinear modal system) is frequently associated with the so-called multimodal method whose ideas and mathematical procedures originated in the 1960s through 1980s from G. Narimanov, I. Lukovsky, L. Dokuchaev, H. Abramson, R. Hutton, and J. Miles, and later on, in the 1990s through 2010s it was extensively developed by O. Faltinsen, A. Timokha, T. Ikeda, M. La Rocca, J. Love, and others.

The authors started working on the present material bearing in mind an extended chapter in a CRC Press/Taylor & Francis Group book. The chapter had to report both the most recent authors' results on the damped resonant sloshing in an upright circular cylindrical tank performing an orbital periodic motion and the associated mathematical and physical fundamentals. However, after composing several sections, we realised that the single-chapter format is not a suitable presentation format if we wish the chapter to be understandable and useful for all potentially interested mathematicians and engineers, especially for newbies in sloshing. This is, basically, because of a rather low page limit that makes it impossible to outline all the needed mathematical and physical details—implementing the multimodal method requires knowing numerous mathematical and physical results, which are spread out between diverse journal papers. Furthermore, proceeding analytically with the multimodal method strongly depends on the tank shape, that is, each tank geometry requires a dedicated consideration, yielding a specialised version of the multimodal theory, and, therefore, one should include even more of those details to clarify how the method works when dealing with the chosen tank geometry. The practically important containers deserve specialised books.

Instead of writing the chapter, we concentrated on a book devoted to sloshing in an upright circular cylindrical shape with a finite liquid depth and asked Professor Hemen Dutta whether we could prepare and publish such a book, where the authors' results, mathematical fundamentals, and specific details of the multimodal method for the considered tank shape are reported. We felt the book should also contain tables with hydrodynamic coefficients, which facilitate engineers who wish to use the existing modal equations but do not want to know mathematical aspects of the multimodal method. His answer was positive.

The authors thank Professors Ivan Lukovsky (Kyiv, Ukraine) and Odd Faltinsen (Trondheim, Norway) for fruitful collaboration before and during preparation of the book. The second author also acknowledges the financial support of the Centre of Autonomous Marine Operations and Systems (AMOS) whose main sponsor is the Norwegian Research Council (Project number 223254-AMOS).

Trondheim, Kyiv
February 2020

Ihor Raynovskyy
Alexander Timokha

Authors

Ihor Raynovskyy graduated from the Lviv Physics and Mathematics Lyceum, Ukraine, in 2009, where he used to be a part of the Olympiad teams in both mathematics and physics. In 2013, he obtained a BS in Actuarial and Financial Mathematics, and, later on, an MS in Mathematics (2015) from the Kyiv National Taras Shevchenko University, Ukraine. He defended his PhD (supervisor Professor A.N. Timokha) in 2019 at the Institute of Mathematics of the National Academy of Sciences of Ukraine. His thesis was devoted to applied mathematical aspects of the damped liquid sloshing. Since 2018, he has occupied the researcher position at the Institute of Mathematics. Dr. Raynovskyy has authored seven journal papers and participated in five international conferences.

Alexander Timokha obtained his PhD (1988) in fluid mechanics from the Kyiv Taras Schevchenko State University, USSR, and, later on, he received his doctoral degree (Habilitation) in physics and mathematics (mathematical physics, 1993) from the Institute of Mathematics of the National Academy of Sciences of Ukraine where he works as a full professor in applied mathematics (since 1994) and head of the department of Mathematical Problems of Mechanics and Control Theory (since 2017). In 2015, the National Academy of Sciences of Ukraine elected him Corresponding Member. Professor Timokha is the Senior NATO (1998-1999) and Alexander-von-Humboldt (2003-2004, Germany) Fellow. He obtained many national and international awards including the Presidential (of Ukraine) Early Career Researcher Award (in both 1995 and 1996), Petryshyn's prize for the best work in nonlinear analysis (USA, 1994), the prize named after M. Krylov (2001), and the State Award in Science and Technology (2012). Dr. Timokha has been principal investigator of numerous national and international projects, including those funded by INTAS (EU, 1994–1995) and Deutsche Forschungesgemeinschaft (Germany, 1997-2010). Professor Timokha is a member of the American Mathematical Society (USA, since 1994), Gesellschaft für Angewandte Mathematik und Mechanik (Germany, invited member, since 1995) and EUROMECH. Currently, he is also an editorial board member of the *Ukrainian Mathematical Journal, Mathematical Problems in Engineering*, and *Ocean Systems Engineering*. In 2004-2012, Dr. Timokha was the visiting professor at the Centre of Excellence "Centre for Ship and Ocean Structures", Norwegian University of Science and Technology (NTNU, Norway). Since 2013, he occupies the permanent research professor position at the Centre of Excellence "Autonomous

Marine Operations and Systems" (NTNU, Norway). His interests lie in free-surface wave problems. Professor Timokha has authored more than 200 journal papers and 6 books including the well-known bestselling *Sloshing*, written with Professor Odd M. Faltinsen and published by Cambridge University Press in 2009.

1 Introduction

A rigid tank carrying a liquid with a free surface, is, from physical, mechanical, and mathematical points of view, a *hybrid* mechanical system. This hybrid system implies that the rigid tank moves, normally, with six degrees of freedom governed by a system of ordinary differential equations (dynamic equations of the rigid body), but the liquid sloshing dynamics is described by a free-surface (boundary) problem.

The multimodal method is a most representative and popular analytical approach used to study liquid sloshing dynamics. Its idea consists of replacing, in a rigorous mathematical way (not phenomenological as when adopting the so-called equivalent mechanical systems), the free-surface (sloshing) problem by a discrete mechanical system (a system of ordinary differential equations) with an infinite number of degrees of freedom. As it is normally accepted in the analytical mechanics, the discrete system is a product of variational principles for the original continuum problem.

The multimodal method introduces the *hydrodynamic generalised coordinates*, which are normally associated with contribution (perturbation) of the standing (Stokes) waves (natural sloshing modes). Furthermore, as we stated above, the method adopts an appropriate variational principle, the so-called modal (Fourier) representation of the free surface, whose time-depending coefficients are the hydrodynamic generalised coordinates and, as a result of implementing the variational technique, the method derives a (modal) system of ordinary differential equations coupling the hydrodynamic generalised coordinates and velocities. The modal system is, in fact, the Euler-Lagrange type equation of the second kind for the considered hydrodynamic system. It is fully equivalent to the original free-surface problem. Analysing the modal system makes it possible to study, in an analytical way, resonant sloshing regimes, their stability, secondary resonances, chaos (irregular waves), and other sloshing phenomena even for the case when implementing computational fluid dynamics (CFD) is questionable and/or inefficient. The latter is especially for steady-state wave flows occurring on the long time scale after all the transient sloshing components disappear (to specify the steady-state waves and study their stability by using a CFD method, exhaustive search through all admissible initial scenarios is required). No doubts, using the modal system has many theoretical advantages in simulating the coupled liquid-rigid container dynamics.

The primary goal of this book is to demonstrate most important mathematical and mechanical details of the multimodal method for an upright circular cylindrical tank. This includes the Bateman-Luke variational formalism for the original free-surface (sloshing) problem, the Miles-Lukovsky modal equations, which are fully equivalent to the original free-surface problem and play the role of the Euler-Lagrange equation for the hydrodynamic system, a series of existing approximate modal systems (linear and weakly-nonlinear), as well as analytical approaches to the weakly-nonlinear modal systems; over here, a focus is on constructing the steady-state (periodic) solutions, examining their stability, and classifying the associated resonant wave regimes.

1.1 ORIGINS OF THE MULTIMODAL METHOD

The authors guess that the termini *"multimodal method"* was first proposed in the pioneering journal paper [54] by Faltinsen, Rognebakke, Lukovsky, and Timokha. However, the method originated forty years before, in the 1950s and 1960s, when applied mathematicians and engineers were seriously challenged by aircraft, spacecraft, and marine applications. An enthusiastic atmosphere of these years is described in the memoir article by Abramson [3] who headed a space NASA program [2, 37, 73, 95, 213, 219, 220]. Analogous spacecraft programs existed in the Soviet Union. However, their engineering results were, basically, not published in the open literature; exceptions are the applied mathematical books [1, 72, 107, 159, 160, 169, 170, 172, 194]. Experimental studies by the Soviet researchers were done in Moscow, Kiev [113, 147], Dnipropetrovsk [18], and Tomsk [19, 21].

1.1.1 LINEAR SLOSHING

Publications of the 1950s and 1960s primarily concentrated on *theoretical* studies of the linear (small-amplitude) sloshing, observing and measuring, experimentally, *nonlinear* sloshing phenomena, as well as designing slosh-suppressing devices.

The *linear multimodal method* (see its original [178, 185, 193], canonical [72, 148, 159, 226], and modern [62, 174] descriptions) was proposed in the 1950s. The authors did not associate their results with variational formalism of analytical mechanics. However, what they used to derive is, in fact, the Euler-Lagrange equation of the second kind for the linear sloshing problem as if they were employing the Lagrangian variational principle. A mathematical link between the Lagrangian formalism in the linear sloshing problem and the corresponding linear modal equations was established in [148]. The book also considers the coupled rigid body-liquid dynamics by using the linear multimodal method and shows that the Lagrangian formalism leads to the Euler-Lagrange type equation. Practically, the multimodal method reduces the linear sloshing (boundary-value) problem to an infinite set of linear oscillators, which is called, all together, the linear modal system. The inhomogeneous terms in the right-hand side of the linear modal system are functions of the six generalised coordinates (degrees of freedom) of the rigid body. Analogously, the differential (governing) equations of the rigid body contain in their right-hand side a linear expression by the hydrodynamic generalised coordinates and their time-derivatives. The latter right-hand side is associated with the resulting hydrodynamic force and moment.

The linear multimodal method takes into account that the liquid (inviscid incompressible with irrotational flows) sloshing is a conservative mechanical system. The method implementation needs to know, *a priori*, the *natural sloshing* modes, $\varphi_n(x,y,z)$, and frequencies, σ_n, as well as the so-called *Stokes-Joukowski potentials*, three time-independent harmonic functions $\Omega_{0i}(x,y,z)$, $i=1,2,3$, defined in the mean free domain Q_0. Coupling the linear modal system and the dynamic equations of the carrying rigid container can be facilitated by utilising the linearised *Lukovsky formulas*, which express the resulting hydrodynamic force and moment in terms of the hydrodynamic generalised coordinates. The hydrodynamic coefficients in both the linear modal system and the Lukovsky formulas are integrals over Ω_{0i}, φ_n, and their spatial derivatives. This means that, when having known Ω_{0i} and φ_n, one can immediately get the governing equations for the coupled liquid-rigid body dynamics and, further, using standard approaches to linear ordinary differential equations, find analytical solutions of the linear sloshing problem (see, chapter 5 of [62] and [63,105,106,239]) and even describe the *linear coupled dynamics*. Details for an upright circular cylindrical tank are reported in Chapter 3.

One should remember that the Stokes-Joukowski potentials are needed to describe hydrodynamic loads caused by angular tank motions. The loads are not identical to those by a 'frozen' liquid. The potentials come from the Neumann boundary value problem in Q_0. They were first derived by Nikolay Joukowski (1885) [102] who examined a spatially-moving rigid body with a cavity completely filled with an ideal incompressible liquid. Exact analytical Ω_{0i} are a rare exception [62]. However, these exist for upright tanks with circular, annular, and rectangular cross sections.

The natural sloshing modes are eigenfunctions of a special spectral boundary problem in the mean (hydrostatic) liquid domain Q_0. The spectral parameter κ appears in the mean free-surface (Σ_0) boundary condition so that $\sigma_n = \sqrt{\kappa_n g}$ are the natural circular sloshing frequencies (g is the gravity acceleration). The traces $\varphi_n|_{\Sigma_0}$ define the standing wave patterns. For the upright circular base container, the analytical solution of the spectral boundary problem was first constructed by Mikhail Ostrogradsky. His manuscript [186] was submitted to the Paris Academy of Sciences in 1826 and, later on, revisited by Poisson and Rayleigh [38] for other tank shapes. The rigorous mathematical theory on the natural sloshing modes and frequencies was created in the 60's (chapter VI in [72] and [42,174]). It states, in particular, that (i) the spectrum uniquely consists of positive eigenvalues, κ_n, with the only limiting point at infinity (in contrast to the external water waves problem, which yields a continuous spectrum), (ii) $\{\varphi_n|_{\Sigma_0}\}$ constitute, together with a non-zero constant, a functional basis on Σ_0 in the mean-square metrics $L_2(\Sigma_0)$. The first fact is important for understanding why the Kordeweg-de-Vries, Boussinesq, etc.

partial differential equations (sea waves) and ordinary differential (modal) equations (contained liquid) follow from the same free-surface problem but have different mathematical structures and solution types. The second fact is a foundation stone for introducing the hydrodynamic generalised coordinates. In the 1960s and 1970s, S. Krein [104, 110] (see, also [31, 32]) generalised these results on the natural sloshing modes to a *viscous incompressible liquid*, but N. Kopachevskii — to the *capillary liquid* (Part II in [176] and [110, 111]).

In the 1950s through the 1980s, diverse analytically approximate natural sloshing modes and Stokes-Joukowski's potentials were constructed (see, a collection in [148] and semi-analytical solutions for circular/spherical tanks in [26]). Because of the volume conservation condition, these solutions, normally, resulted from implementing the Trefftz method, which employs a harmonic functional basis, in particular, harmonic polynomials [148] whose completeness in the star-shaped domains was proven in [228, 229]. The Trefftz approximations become especially efficient and provide a uniform convergence when the corner-point singularity [72, 108, 109, 236, 237] is accounted for [63, 69]. Nowadays, the harmonic polynomials are extensively used in the harmonic polynomial cell (HPC) method [206, 207]. Growing computer capabilities of the 1990s to 2000s made finding φ_n and Ω_{0i} no problem. An exception may be non-clean tanks equipped with baffles, screens, and other slosh-suppressing devices that leads to a strongly singular behaviour of $\nabla\varphi_n$ at their sharp edges [64, 83].

1.1.2 NONLINEAR SLOSHING

Theoretical studies of the nonlinear (resonant) sloshing were initiated in the 1950s and 1960s by Penny and Price [188], Moiseev [171], Narimanov [179], and Perko [173, 189]. By using a perturbation technique [188], Moiseev [171] showed how to construct an *asymptotic steady-state (periodic)* solution of the nonlinear (free-surface) sloshing problem in a rigid tank. He assumed that either the tank performs a prescribed small-amplitude harmonic motion with the forcing frequency σ close to the lowest natural sloshing frequency σ_1 or there is an extra $2\pi/\sigma$-periodic resonant pressure acting on the free surface. The analysis suggested a finite liquid depth of an ideal incompressible liquid with irrotational flows. Moiseev proved that, if the nondimensional forcing amplitude (scaled by the tank breadth) is a small parameter $\epsilon \ll 1$, contribution of the primary excited natural sloshing mode is of the order $O(\epsilon^{1/3})$ and the matching resonant asymptotics (the so-called Moiseev detuning) is

$$(\sigma^2 - \sigma_1^2)/\sigma_1^2 = O(\epsilon^{2/3}). \tag{1.1}$$

The Moiseev theory implicitly assumed that there is no the so-called *secondary resonance* [25, 60–62, 238]. It was originally realised for the two-dimensional rectangular tank [47, 182]. Other tank shapes were considered in [13, 95, 156, 211]. Getting the Moiseev solution deals with huge and tedious derivations.

In the 1980s, when looking for an *almost-periodic* resonant solution, Miles [162, 163] generalised Moiseev's results by deriving the so-called *Miles equations*, which govern a slow-time variation of dominant, $O(\epsilon^{1/3})$, amplitudes of the primary excited hydrodynamic generalised coordinates. He considered a horizontal harmonic excitation of an upright circular cylindrical tank and adopted the Moiseev detuning (1.1). Separation of the fast and slow time scales was done in the *Bateman-Luke variational* formulation [10, 89, 129]. The Miles equations were later derived for an upright rectangular tank. Using these equations is a rather popular approach in applied mathematical studies of the almost-periodic resonant sloshing, in particular, for clarifying the chaos in this hydrodynamic system [94,162,163]. Both horizontal and vertical (Faraday waves) harmonic excitations were considered in [76,77,90,91,94,165–167]. The works [112, 208] utilised the Miles equations for studying the 'rigid tank-contained liquid' system with a limited power supply forcing.

By using a perturbation technique, Narimanov [179] derived a version of *weakly-nonlinear* modal systems with multiple degrees of freedom. Narimanov did not know Moiseev's results, but he postulated asymptotic relations between the hydrodynamic generalised coordinates and velocities as if these might follow from the Moiseev asymptotic solution. The original derivations in [179] (the same in [214–217]) contained algebraic errors, which were corrected by Lukovsky [139, 143, 147, 180]. Narimanov's perturbation technique also causes huge and tedious derivations dramatically increasing with increasing the number of the hydrodynamic generalised coordinates in the derived modal system. This explains why all existing Narimanov-type modal systems couple only from two to five hydrodynamic generalised coordinates. These systems were derived for upright tanks of circular, annular, and rectangular cross sections, conical and spherical tanks, as well as for upright circular cylindrical tanks with rigid-ring baffles [84, 139, 143, 180]. Nowadays, the Narimanov method is rarely employed; instead, variational versions of the multimodal method [130–132] are used, based, normally, on the Bateman-Luke variational formalism [10, 62, 89, 129, 147].

1.1.3 PERKO'S METHOD

In the 1960s through the 1980s, simulations of the nonlinear sloshing were based on the Galerkin projective scheme, finite difference, e.g., maker-and-cell, and finite element methods [6, 43, 44, 48, 175, 177, 184, 195, 209, 218]. The multimodal method was also considered as a numerical tool in the works by Perko [173, 189]. In the 1970s and 1980s, the Perko method (sometimes it is named the spectral method) was used, with diverse modifications, by Chakhlov [20, 40] and Limarchenko [118–124] who combined it with the Lagrange variational formalism and a specific perturbation technique.

1.2 THE NONLINEAR MODAL MODELLING

In the 1990s to 2000s, studies on nonlinear sloshing split into *computational* and *analytical* directions. The same happened in the sea waves theory, which is nowadays investigated, almost independently, by utilising computational fluid dynamics (CFD) methods and employing approximate analytical models (exemplified by the Kordeweg-de-Vries and Boussinesq equations [187]). The analytical models are typically not used in engineering computations but remain useful for quantifying and classifying the nonlinear resonant steady-state waves and their stability, detecting solitons, and studying other specific surface-wave phenomena.

A success of CFD was initially associated with the FLOW-3D, which was the most popular commercial Navier-Stokes solver of the 1990s and 2000s (Solaas [210] discussed its advantages and drawbacks). Further, the volume of fluid (VoF), smoothed partitions hydromechanics (SPH), and their modifications provided, based on parallel computation algorithms, very accurate, efficient, and robust simulations ([28] outlines the CFD of the 1980s and 1990s; more recent achievements are reported in [199] and in introductions of [27, 45, 87, 221, 231, 240, 241]). Contemporary CFD solvers are typically based on viscous and fully-nonlinear physical statement. By conducting simulations with different initial scenarios (conditions), they are able to describe the specific free-surface phenomena such as fragmentation, wave breaking, overturning, roof and wall impacts, and flip-through. The aforementioned phenomena cannot be accurately predicted within the framework of physical (mathematical) models adopted by analytical methods. These models are, normally weakly-nonlinear, assume an ideal liquid with irrotational flows, and smooth instant free-surface patterns.

Because the Perko method pursues direct numerical simulations but employs simplified physical (mathematical) models, it lost the competition to both the CFD (in an accurate prediction of hydrodynamic loads) and the analytical sloshing. This explains why the method is rarely used today, restricted to a few fully-nonlinear simulations by the Miles-Lukovsky system for a rectangular tank in [114, 205] and the weakly-nonlinear simulations in [50, 59, 121–123, 125–127]. Another drawback of the Perko-type simulations is that these are unrealistically *stiff* so that an artificial damping must be incorporated to kill away the rising parasitic higher harmonics [50, 59]. Performing those simulations may be justified for screen-equipped tanks when the damping is relatively high and artificial damping terms are not needed [125, 126, 128].

Modern analytical approaches to the nonlinear sloshing are frequently associated with the multimodal method [130–132] based on the Bateman-Luke variational formalism [10, 62, 89, 129, 147], which derives, in a natural way, both dynamic and kinematic relations of the problem [62, 143, 204, 234, 235]. The corresponding variational multimodal method was proposed in 1976 by Miles and Lukovsky [143, 161], who derived, independently, the fully nonlinear modal system coupling the hydrodynamic generalised coordinates and velocities; they considered sloshing in an upright tank performing a prescribed translational motion. Lukovsky re-derived the system for an arbitrary rigid tank motion [143]. He also proposed a *non-conformal mapping technique* to get the modal equations for tanks with nonvertical walls [67, 78, 136, 139, 146] and derived the so-called *Lukovsky formulas* [143, 147] (Chapter 7 of [62] gives an alternative derivation), which express the resulting hydrodynamic force and moment in terms of the hydrodynamic generalised coordinates. An outline of the analytical results can be found in his recent book [147]. His works [142, 143, 147] present a generalisation of the Bateman-Luke variational formalism for the coupled 'rigid tank-contained liquid' mechanical system [142, 143]. Using the Reynolds hypothesis of viscous flows, Lukovsky demonstrated how to incorporate the linear damping into the variational formulation [144, 147] and derived the corresponding nonlinear modal equations.

Adopting specific asymptotic relationships between the hydrodynamic generalised coordinates and velocities reduces the Miles-Lukovsky modal system to what is today called the *weakly-nonlinear* multimodal system [62, 92, 133, 137, 147]. Both the Miles-Lukovsky and weakly-nonlinear multimodal systems require the analytically approximate natural sloshing modes, φ_n, and the Stokes-Joukowski potentials, Ω_{0i}, which must be analytically continuable over the mean free surface Σ_0. These and other limitations are discussed in Chapter 7 of [62] and in [67, 92, 137].

1.3 NONLINEAR MODAL SYSTEMS AND STEADY-STATE ANALYSIS

1.3.1 UPRIGHT TANKS

1.3.1.1 Two-dimensional sloshing in rectangular tanks

The *two-dimensional* weakly-nonlinear resonant sloshing in a clean rectangular tank with a finite liquid depth was studied by using diverse weakly-nonlinear multimodal systems in [54, 60, 74, 92, 93, 97, 98, 101]. The forcing amplitude was small relative to the tank width, of the order $O(\epsilon) \ll 1$, and the forcing frequency σ was close to the lowest natural sloshing frequency σ_1. The nonlinear modal system in [54] was based on the Narimanov-Moiseev asymptotics and (1.1). Using the system, transient and steady-state resonant predictions were obtained and validated for both the prescribed harmonic tank excitation [54] and the coupled dynamics of a rigid floating tank in an external seaway [116, 117, 202].

The steady-state resonant sloshing caused by the prescribed harmonic tank motions is characterised by the Duffing-type response with the theoretically soft-spring type behaviour as the *depth-to-breadth ratio* $> 0.3368...$ and the hard-spring type behaviour for $< 0.3368...$ so that the theoretical *non-dimensional critical depth* is equal to 0.3368... [62, 232]. A purely mathematical analysis of the weakly-nonlinear multimodal system was reported in [74, 92, 93].

The Narimanov-Moiseev–type weakly-nonlinear multimodal system [54] becomes irrelevant with increasing the forcing amplitude, at the critical and small (shallow) liquid depths due to the *secondary resonance* in the mechanical system, which suggests that $n\sigma \approx \sigma_n$ for the natural sloshing mode with

$n \neq 1$. The secondary resonance leads to a nonlinearity-driven amplification of the $n\sigma$ Fourier harmonics and an energy transfer from primary (at σ_1) to secondary (σ_n) excited natural sloshing mode. Handling the secondary resonance for the finite liquid depth requires the so-called *adaptive* modal systems [60, 93] suggesting a few extra dominant (secondary excited) hydrodynamic generalised coordinates of the order $O(\epsilon^{1/3})$, for which the dominant higher harmonics contribution is theoretically predicted ($n\sigma \approx \sigma_n$). The adaptive multimodal method was validated by comparison with experiments in [59–61]. When the secondary resonance occurs, the response curves demonstrate several peaks in the primary resonance zone. Existing adaptive modal systems and their structure are extensively discussed in Chapter 8 of [62]. Based on the adaptive modal modelling, [93] showed that the *critical depth* is a function of the forcing amplitude so that the aforementioned 0.3368... is the limiting value as the nondimensional forcing amplitude tends to zero. This fact clarifies why experimental value of the critical depth is estimated at 0.28 in the classical experiments [75]. The adaptive modal systems [97, 98, 101] employ a sophisticated asymptotic ordering to handle the secondary resonance phenomena.

The commensurate/almost-commensurate sloshing eigenspectrum for *shallow/small depths* leads to hydrodynamic jumps [230]/multiple secondary resonances [61]. Deriving the weakly-nonlinear multimodal system [61] for the small liquid depth in a rectangular tank needs the Boussinesq–type fourth-order asymptotic ordering, which combines Moiseev's and Kordeweg-de-Vries' asymptotics [181–183]; all the hydrodynamic generalised coordinates and the nondimensional liquid depth are then of the same asymptotic order $O(\epsilon^{1/4})$, where, as usually, $O(\epsilon)$ measures the forcing amplitude. By truncating this infinite-dimensional system and incorporating the linear damping terms (reflecting dissipation caused by the laminar viscous boundary layer and the bulk viscosity; see Chapter 6 in [62] and [30, 103, 157, 168, 227]), the paper [61] conducts a numerical analysis, which provides a good agreement with experimental measurements in [35, 36, 61] for both steady-state and transient waves. As identified in Chester's experiments [35, 36, 61], the theoretical steady-state response curves in [5, 61] possess a fingers-type shape with many peaks in the primary resonance zone ($\sigma \approx \sigma_1$). Increasing the excitation amplitude and/or decreasing the liquid depth make the weakly-nonlinear multimodal system [61] physically inapplicable due to breaking and overturning waves, bores, and free-surface fragmentation, which yield, all together, the enormous damping. A detailed experimental classification of the shallow liquid sloshing (four different wave types were specified) is reported in Chapter 8 of [62]. Damping due to breaking and overturning waves, bores, and free-surface fragmentation is similar to that for the *roof impact*, whose effect was included into the multimodal systems of [54, 60] in [53] by using the Wagner theory.

Damping caused by the flow separation through a *perforated screen* was included into the weakly-nonlinear multimodal systems [49–51, 125] for two-dimensional sloshing in a rectangular tank with a finite liquid depth. For the smaller solidity ratio of the screen, $0 < Sn < 0.5$, and the relatively small forcing, utilising the pressure drop condition [17] yields a specific integral damping term in the modal systems [49, 125]. The modified multimodal systems show a satisfactory agreement with experiments. The higher solidity ratio, $0.5 < Sn < 1$, modifies the natural sloshing modes and frequencies [64] and, thereby, both linear [51] and nonlinear [50] multimodal systems changed their analytical structure. The secondary resonance became then evident so that the multi-peak response curves in the primary resonant zone differ from those for the clean tank. Other weakly-nonlinear multimodal systems for various tanks with perforated screens were derived, studied, and validated in [88, 125–128].

1.3.1.2 Three-dimensional sloshing in rectangular tanks

A generalisation of [54] to the *three-dimensional rectangular* tank was done in [55]. A focus has been on the nearly-square cross section leading to the degenerating natural sloshing modes for the lowest natural frequency σ_1. The corresponding Narimanov-Moiseev–type multimodal system in [55, 56] has 9 degrees of freedom with the two dominant $O(\epsilon^{1/3})$-order hydrodynamic generalised coordinates. The system provided an accurate classification of the steady-state waves (planar, diagonal, nearly-diagonal, and swirling) for both longitudinal and diagonal harmonic tank excitations [55–57]. The asymptotic periodic solution of the multimodal system possesses an amazing bifurcation diagram, especially, when the cross-section aspect ratio changes from the unit [23, 24, 58]. A good quantitative agreement with experiments was shown. On the other hand, the theoretical transient and steady-state wave response was not quantitatively supported by experiments due to the secondary resonance, which becomes especially evident for swirling. Accounting for the secondary resonance effect led to adaptive modal systems [57, 59]. Computational experiments with the damped adaptive modal systems and different initial scenarios in [57, 59] established specific numerical solutions, which are periodic by the dominant generalised coordinates but the higher-order generalised coordinates demonstrate an irregular (chaotic) behaviour because of a kind of instability. The numerical solutions co-exist with the asymptotic periodic solutions of the modal systems. Experiments support the latter fact. Specifically, the liquid damping is satisfactorily predicted only for the lowest natural sloshing modes (Chapter 6 of [62] and [59, 103, 157, 168]). The damping of the higher natural sloshing modes is strongly affected by the randomly-occurring free-surface fragmentation observed in [56, 57, 59].

1.3.1.3 Annular and circular tanks

In the 1980s, Lukovsky [134, 142, 143] derived a five-dimensional weakly-nonlinear modal system approximating the resonant sloshing in an *upright circular cylindrical tank*, constructed its asymptotic periodic solutions (described planar and swirling wave modes) for the longitudinal harmonic forcing, studied their stability by the first Lyapunov method, established irregular (chaos, modulated, etc.) waves in a certain frequency range, and validated these steady-state wave results by experiments [4]. The paper [99] re-derived this Lukovsky system, re-classified the steady-state wave regimes, and conducted illustrative Runge-Kutta's simulations (as in [143, 147]). This system and constructed periodic solutions were revisited in [81] and Chapter 9 of [62]. Assuming the fifth-order asymptotic ordering ($O(\epsilon^{1/5})$ for the two dominant hydrodynamic generalised coordinates) gives a negligible correction to the steady-state analysis [153].

The Narimanov-Moiseev's asymptotics requires an infinite set of the second- and third-order hydrodynamic generalised coordinates in the weakly-nonlinear multimodal systems for *axisymmetric tanks* [52, 67, 137]. For an upright circular cylindrical tank, such a 'mathematically complete' infinite-dimensional multimodal system was derived and analysed in [52, 133]. The infinite set of the higher-order hydrodynamic generalised coordinates did not qualitatively influence the periodic steady-state solutions due to the longitudinal forcing except for several isolated liquid depths and forcing frequencies when the secondary resonance matters (the critical liquid depths are listed in [25, 52]).

Lukovsky derived and studied the corresponding five-dimensional multimodal system [143, 190, 191] for the resonant sloshing in an *upright annular* cylindrical tank. The system was re-derived and modified by adding extra third-order hydrodynamic generalised coordinates in [223], where new experimental model tests were also conducted. Comparing the experimental and theoretical response curves [223] showed a satisfactory agreement for the planar wave regime, but even though the speculative damping ratios were included to fit the experimental data, a discrepancy was still evident for the swirling wave mode. An attempt to derive a Narimanov-Moiseev–type multimodal system for a non-central pile position was done in [222]. Sloshing in an upright *compartment* tank of circular and annular cross sections was studied in [154, 155] by using the analytical scheme from [143, 147].

The paper [52] derived an infinite-dimensional system of the Narimanov-Moiseev–type equations for an annular tank. But the authors did not account for the liquid damping, which may matter. Timokha and Raynovskyy [198, 224, 225] included the linear damping terms into the multimodal analysis. Their results will be reported in the present book.

1.3.2 NONCONFORMAL MAPPING TECHNIQUE

For tanks with *non-vertical walls*, there are no exact analytical natural sloshing modes, and the single-valued (normal) representation of the free surface is not possible. The later problem could be resolved by utilising the so-called *non-conformal mapping technique* proposed by Lukovsky [138, 139, 141, 180] in 1975. The technique was combined with the Narimanov asymptotic scheme in [139, 180], used in the Perko simulations [122–124], and was incorporated into the variational multimodal approach [67, 143, 147]. The corresponding multidimensional weakly-nonlinear modal systems were derived for *conical* and *spherical* tanks. The main difficulty for a further expansion onto other tank shapes remains the absence of the required analytically approximate natural sloshing modes, which exactly satisfy the Laplace equation and the slip conditions on the non-vertical walls as well as allow for an *analytical continuation* over the mean free surface (the limitations are discussed in [137,147] and Chapter 9 of [62]).

In [12, 13, 39], the flat mean free surface in a circular conical tank was replaced by a spherical cap and, thereby, an exact analytical solution of the spectral sloshing problem was constructed in terms of spherical functions. The corresponding multidimensional weakly-nonlinear modal system was derived in [149, 152]. The latter result was step-by-step improved in [78, 80, 85, 86, 135, 143, 145, 212] (see, also references therein). Analytically approximate natural sloshing modes without the aforementioned replacement were constructed in [80, 86, 135]. Utilising these modes, the Narimanov-Moiseev-type five-dimensional nonlinear multimodal systems for the *non-truncated and truncated circular V-conical tanks* were derived and studied in [78, 86]. The systems contain specific nonlinear terms expressing the so-called geometric nonlinearity, but the steady-state sloshing analysis for the longitudinal forcing leads to qualitative-similar results to those for an upright circular cylinder.

The work [82] focused on the moving nodal lines that generalise [81]. An emphasis was also on the secondary resonance expectations versus the conical tank geometry. The secondary resonance analysis in [78, 86] and a comparison of the theoretical results with experiments [29, 78, 85, 158] showed that the secondary resonance indeed matters. The works [150, 212] summarise the results on the multimodal method for circular conical tanks including derivation of the infinite-dimensional modal system.

The required analytically approximate natural sloshing modes for the *circular* and *spherical* tanks were constructed in [7, 8, 65, 66]. Based on those solutions from [8, 65], an infinite-dimensional Narimanov-Moiseev-type multimodal system (generalisation of [133]) was derived in [67, 68], and their steady-state (periodic) solutions were constructed and analysed. The latter results were supported by experiments [213] for the depth-to-radius ratios ≤ 0.5. However, secondary resonances and an experimentally observed fragmentation of the free surface made this weakly-nonlinear multimodal system inapplicable for the higher tank fillings as well as with increasing the forcing amplitude. The Narimanov-Moiseev-type infinite-dimensional multimodal system for a circular tank was constructed but not analysed in [33]. The circular tank shape causes multiple secondary resonances for almost all tank fillings [18, 63].

1.4 CHALLENGES OF THE MULTIMODAL MODELLING

Main challenges of the nonlinear multimodal modelling are associated with its generalisation onto new practically-important sloshing problems (Chapter 1 of [62] and [34, 147]), improvement of existing nonlinear multimodal systems by letting them account for the damping due to the free-surface phenomena including the wave breaking, as well as modification of analytical approaches to analysing the multidimensional modal systems [74, 92, 93].

An obvious challenge is how to deal with tanks of complex shape. As remarked above, this requires constructing the analytically-approximate natural sloshing modes of special kind and, when applying the Lukovsky nonconformal mapping technique, developing a tensor algebra related to the corresponding curvilinear coordinates [147]. Because derivations of the nonlinear multimodal systems for the complex tank shapes are rather difficult and tedious, a challenge may be the coding of a computer algebra, which facilitates computer-based derivations as it was done for an upright circular cylindrical tank in [52, 79].

Two important physical challenges are an adequate account for damping and surface tension. Chapter 6 of [62] reviews analytical methods, which could be used for incorporating the damping-related terms in the existing nonlinear multimodal systems. However, these were only developed for the laminar viscous boundary layer, the bulk viscosity, and, partly, for flow separation at perforated screens [50, 51, 125]. Accounting for damping due to flow separation around baffles and piles should be the next goal. For the nonlinear sloshing, surface tension requires accounting for the dynamic contact angle effect [15, 16]. It is not clear yet how to introduce the effect into the multimodal method.

Finally, we believe that the multimodal method may help to describe an attractive swirling-induced V-constant rotation of the contained liquid [46], which was observed in the experiments [14, 95, 96, 192, 203]. This topic is associated with an analytical modelling of swirling and mixing phenomena in bioreactors. Many details on that can be found in experimental and theoretical studies of the PhD thesis [200] as well as in [22, 41, 200, 201, 233]. The first step toward solving the problem was made in [71]. But the latter authors did not account for the viscous damping.

2 Differential, variational, and modal statements

Let a liquid partly fill a rigid tank, which moves with 6 degrees of freedom as shown in Figure 2.1. The $Oxyz$ coordinate system is fixed with the tank so that the Oxy plane coincides with the mean (hydrostatic) free surface. Everywhere in this book, the prescribed tank motions are of oscillatory character with a small (with respect to the mean free-surface size R_0) magnitude. The liquid-tank mechanical system is located in the gravitational field with the gravity potential

$$U_g = -\boldsymbol{g} \cdot \boldsymbol{r}' = -\boldsymbol{g} \cdot \boldsymbol{r} - \boldsymbol{g} \cdot \boldsymbol{r}'_0, \qquad (2.1)$$

where $\boldsymbol{g}$ is the gravity acceleration, $\boldsymbol{r} = x\boldsymbol{e}_1 + y\boldsymbol{e}_2 + z\boldsymbol{e}_3$ is the radius-vector in the tank-fixed coordinate system ($\boldsymbol{e}_i$ are the coordinate units), and $\boldsymbol{r}'_0$ is the radius-vector of the origin O relative to an inertial coordinate system $O'x'y'z'$. Henceforth, except in section 2.2.2, the book assumes:

1. The contained liquid is incompressible, perfect, and its flows are irrotational. It does not contain gas cavities (bubbles, gas pockets, etc.).
2. Surface tension is neglected that requires such a characteristic size of the mean free surface (here, this is the radius R_0 for the upright circular container) that the Bond number, Bo= $gR_0^2\rho/T_s$ (ρ is the liquid density and T_s is the surface tension) is large. Analysing the liquid sloshing in the microgravity, [176] recognised that the Bond number $\approx$ 100 matters for the capillary meniscus (the hydrostatic free surface is normally far from planar at the tank wall), but both the free-surface elevation slightly away from the tank walls and the resulting hydrodynamic force and moment are not affected by the free-surface capillarity. Surface tension becomes fully negligible as $1000 \lesssim$ Bo.
3. The thickness of the viscous boundary layer on the wetted tank surface is less than the surface wave magnitude so that the contained liquid indeed behaves as ideal and with irrotational flows away from the (laminar) boundary layer, but generally speaking, the associated viscous damping cannot be neglected.

2.1 FREE-SURFACE PROBLEM

Here and thereafter, $Q(t)$ is the time-dependent liquid domain, $\Sigma(t)$ is the free surface, $S(t)$ is the wetted tank surface, Q_0 is the unperturbed (hydrostatic) liquid domain, Σ_0 is the mean free surface, and S_0 is the mean wetted tank surface. The notations are illustrated in Figure 2.1.

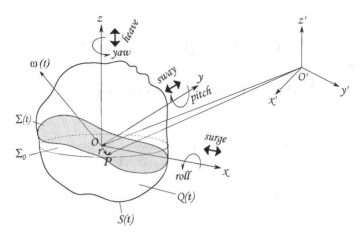

Figure 2.1 A rigid tank partly filled with a liquid. The coordinate system $Oxyz$ is fixed with the tank so that the mean free surface belongs to the Oxy-plane; $O'x'y'z'$ is an inertial coordinate system, which is usually associated with the ground. Relative motions of the $Oxyz$ frame are defined at any time t by the translational velocity $\boldsymbol{v}_O = d\boldsymbol{r}'_0/dt = \dot{\boldsymbol{r}}'_0$ (associated with the origin O) and the instantaneous angular velocity $\boldsymbol{\omega}(t)$ of the rigid body.

2.1.1 GOVERNING EQUATIONS

The governing equation of an ideal incompressible liquid with irrotational flows should be written down in the non-inertial coordinate system $Oxyz$ rigidly fixed with the tank. The absolute velocity field in $Q(t)$ is described by the velocity potential, $\boldsymbol{v} = \nabla \Phi$, which satisfies the continuity equation

$$\nabla \cdot \boldsymbol{v} = 0 \ \Rightarrow \ \nabla^2 \Phi = \Delta \Phi = 0. \tag{2.2}$$

The Laplace equation (2.2) is invariant with respect to the Cartesian coordinate system (inertial or non-inertial) and, therefore,

$$\Delta \Phi \equiv \frac{\partial^2 \Phi}{\partial x^2} + \frac{\partial^2 \Phi}{\partial y^2} + \frac{\partial^2 \Phi}{\partial z^2} = 0 \ \text{in} \ Q(t). \tag{2.3}$$

The Bernoulli equation introduces the pressure field p and takes the form

$$p + \rho \left(\left. \frac{\partial \Phi}{\partial t} \right|_{\text{inertial}} + \frac{1}{2} (\boldsymbol{v})^2 + U_g \right) = C(t) \tag{2.4}$$

in the inertial coordinate system $O'x'y'z'$; here, $C(t)$ is an arbitrary time-dependent function, ρ is the liquid density, and U_g is defined by (2.1). Because $\boldsymbol{r}' = \boldsymbol{r}'_0 + \boldsymbol{r}$ and $\boldsymbol{g} \cdot \boldsymbol{r}'_0$ is only function of the time, one can assume that the gravity potential takes the form

$$U_g = -\boldsymbol{g} \cdot \boldsymbol{r}, \tag{2.5}$$

in the non-inertial coordinate system $Oxyz$.

To write down the Bernoulli equation in this coordinate system, we should investigate $\partial \Phi/\partial t$, which implies the time derivative for a fixed point with coordinates (x', y', z'). Assume that (x, y, z) coincides with (x', y', z') at an instant time t. Difference between (x, y, z) and (x', y', z') at the time $t + \Delta t$, to within the nonlinear terms, is equal to $\boldsymbol{v}_b \Delta t$, where

$$\boldsymbol{v}_b = \boldsymbol{v}_O + \boldsymbol{\omega} \times \boldsymbol{r} \qquad (2.6)$$

(the Euler formula) is the rigid body velocity at a point P defined by the radius-vector $\boldsymbol{r}$; $\boldsymbol{v}_O$ is the velocity of the origin O; and $\boldsymbol{\omega}$ is the instantaneous angular velocity of the rigid body around O. The Taylor series of $\Phi(x, y, z, t + \Delta t)$ at (x', y', z') gives

$$\left.\frac{\partial \Phi}{\partial t}\right|_{\text{in non-inertial } Oxyz} = \lim_{\Delta t \to 0} \frac{\Phi(x, y, z, t + \Delta t) - \Phi(x, y, z, t)}{\Delta t}$$

$$= \lim_{\Delta t \to 0} \frac{\Phi(x', y', z', t + \Delta t) + (\boldsymbol{v}_b \cdot \nabla \Phi)\Delta t - \Phi(x', y', z', t)}{\Delta t} =$$

$$= \left.\frac{\partial \Phi}{\partial t}\right|_{\text{in inertial } O'x'y'z'} + \boldsymbol{v}_b \cdot \nabla \Phi. \qquad (2.7)$$

As a consequence, the Bernoulli equation (2.4) can be rewritten in the non-inertial $Oxyz$ system as follows

$$p + \rho \left(\left.\frac{\partial \Phi}{\partial t}\right|_{\text{non-inertial}} - (\boldsymbol{v}_O + \boldsymbol{\omega} \times \boldsymbol{r}) \cdot \nabla \Phi + \frac{1}{2}(\nabla \Phi)^2 + U_g \right) = C(t). \qquad (2.8)$$

Position of the coordinate system $Oxyz$ implies that $z = 0$ corresponds to the unperturbed (mean) free surface. We assume that the atmospheric pressure in the ullage gas is a constant value equal to p_0. Adopting $C(t)$ to satisfy this condition reduces the Bernoulli equation (2.8) to the following form

$$p - p_0 = -\rho \left(\frac{\partial \Phi}{\partial t} + \frac{1}{2}(\nabla \Phi)^2 - \nabla \Phi \cdot (\boldsymbol{v}_O + \boldsymbol{\omega} \times \boldsymbol{r}) + U_g \right). \qquad (2.9)$$

2.1.2 BOUNDARY CONDITIONS

The boundary conditions on the wetted tank surface $S(t)$ require that the normal velocity of liquid particles is equal to the normal velocity of the rigid tank, i.e.,

$$\boldsymbol{v} \cdot \boldsymbol{n} = (\boldsymbol{v}_O + \boldsymbol{\omega} \times \boldsymbol{r}) \cdot \boldsymbol{n} \quad \text{on} \quad S(t), \qquad (2.10)$$

where $\boldsymbol{n}$ is the outer normal unit to $S(t)$, and $\boldsymbol{v}$ is the absolute velocity. We can rewrite (2.10) as

$$\frac{\partial \Phi}{\partial n} = \boldsymbol{v}_O \cdot \boldsymbol{n} + \boldsymbol{\omega} \cdot (\boldsymbol{r} \times \boldsymbol{n}) \quad \text{on} \quad S(t). \qquad (2.11)$$

Because the pressure p is equal to the atmospheric (ullage) pressure p_0 on the free surface, (2.9) gives the so-called dynamic boundary condition

$$\frac{\partial \Phi}{\partial t} + \frac{1}{2}(\nabla \Phi)^2 - (\boldsymbol{v}_O + \boldsymbol{\omega} \times \boldsymbol{r}) \cdot \nabla \Phi + U_g = 0 \quad \text{on } \Sigma(t). \qquad (2.12)$$

Thus, we have two boundary conditions, one of which is on the wetted tank walls, (2.11), and another on the free surface, (2.12). In addition, we need the kinematic boundary condition on the free surface, which ensures that liquid particles do not leave it. The most general, implicit definition of the free surface is normally associated with the equation

$$Z(x, y, z, t) = 0. \qquad (2.13)$$

Because particles remain on the free surface, the material derivative of Z in the inertial coordinate system $O'x'y'z'$ is zero, that is,

$$0 = \frac{D'Z}{Dt} = \left.\frac{\partial Z}{\partial t}\right|_{O'x'y'z'} + \boldsymbol{v} \cdot \nabla Z = \left.\frac{\partial Z}{\partial t}\right|_{Oxyz} - \boldsymbol{v}_b \cdot \nabla Z + \nabla \Phi \cdot \nabla Z \quad \text{on } \Sigma(t). \quad (2.14)$$

The outer normal to the free surface $\Sigma(t)$ can be written as $\boldsymbol{n} = \nabla Z/|\nabla Z|$, which transforms (2.14) to the kinematic boundary condition

$$\frac{\partial \Phi}{\partial n} = (\boldsymbol{v}_O + \boldsymbol{\omega} \times \boldsymbol{r}) \cdot \boldsymbol{n} - \frac{\partial Z/\partial t}{|\nabla Z|} \quad \text{on } \Sigma(t), \qquad (2.15)$$

where, obviously, the time derivative is calculated in the $Oxyz$-coordinate system.

For upright tanks (the vertical wall is parallel to Oz in the vicinity of the free surface), one can use the single-valued representation of the free surface

$$Z(x, y, z, t) = z - \zeta(x, y, t) = 0. \qquad (2.16)$$

The kinematic boundary condition (2.15) takes then the form

$$\frac{\partial \Phi}{\partial n} = (\boldsymbol{v}_O + \boldsymbol{\omega} \times \boldsymbol{r}) \cdot \boldsymbol{n} + \frac{\partial \zeta/\partial t}{\sqrt{1 + (\nabla \zeta)^2}} \quad \text{on } \Sigma(t). \qquad (2.17)$$

2.1.3 MASS CONSERVATION

The mass of the incompressible liquid must remain constant, that is,

$$M_l = \int_{Q(t)} \rho \, dQ = \rho V_l = \text{const}, \qquad (2.18)$$

where M_l and V_l are the liquid mass and volume, respectively. Equation (2.18) is not automatically satisfied. It defines an admissible class of functions Z (or ζ), which define the instant free-surface shape, that is, (2.18) is the *geometric (holonomic) constraint*.

2.1.4 INITIAL AND PERIODICITY CONDITIONS

Equations (2.3), (2.11), (2.12), (2.14), and (2.18) constitute the free-surface (boundary) problem, which describes the liquid-sloshing dynamics in the $Oxyz$ coordinate system. The problem requires either the initial conditions, or, in the case of periodic vector functions $\boldsymbol{v}_O(t)$ and $\boldsymbol{\omega}(t)$, the periodicity condition.

The initial conditions determine the so-called transient waves, which are excited by the tank motion (by $\boldsymbol{v}_O(t)$ and $\boldsymbol{\omega}(t)$) and initial perturbations of the free surface and/or velocity field. The initial (Cauchy) conditions take the form

$$Z(x,y,z,t_0) = Z_0(x,y,z), \quad \left.\frac{\partial \Phi}{\partial n}\right|_{\Sigma(t_0)} = V_0(x,y,z)|_{\Sigma(t_0)}, \qquad (2.19)$$

where the known Z_0 defines an initial position of the free surface $\Sigma(t_0)$: $Z_0(x,y,z) = 0$ and $V_0(x,y,z)$ defines an initial absolute normal velocity on $\Sigma(t_0)$. Accounting for (2.15), the initial conditions (2.19) can be rewritten in the form

$$Z(x,y,z,t_0) = Z_0(x,y,z), \quad \left.\frac{\partial Z}{\partial t}\right|_{t=t_0} = Z_1(x,y,z)|_{\Sigma(t_0)} = -V_0(x,y,z)|\nabla Z|_{\Sigma(t_0)}. \qquad (2.20)$$

Being equipped with the periodicity condition, the free-surface problem (2.3), (2.11), (2.12), (2.14), (2.18) determines the steady-state wave motions. Stable steady-state waves appear after transients die out. Mathematically, the periodicity condition implies

$$Z(x,y,z,t+T) = Z(x,y,z,t), \qquad (2.21)$$

where $\boldsymbol{v}_O(t+T) = \boldsymbol{v}_O(t)$ and $\boldsymbol{\omega}(t+T) = \boldsymbol{\omega}(t)$, and T is the known forcing period. The condition (2.21) ensures the periodicity of the velocity field, which follows from the Neumann boundary-value problem (2.3), (2.10), (2.17).

2.2 BATEMAN-LUKE VARIATIONAL PRINCIPLE

2.2.1 IRROTATIONAL FLOWS

Theorem 2.1: The Bateman-Luke principle

Smooth solutions of the free-surface boundary problem (2.3), (2.11), (2.12), (2.14), and (2.18) coincide with extrema of the Luke-Bateman action

$$W(Z,\Phi) = \int_{t_1}^{t_2} L\,dt,$$

$$L = \int_{Q(t)} (p-p_0)\,dQ = -\rho \int_{Q(t)} \left[\frac{\partial \Phi}{\partial t} + \frac{1}{2}(\nabla\Phi)^2 - \nabla\Phi\cdot(\boldsymbol{v}_O+\boldsymbol{\omega}\times\boldsymbol{r})+U_g\right]dQ, \qquad (2.22)$$

for the isochronous independent smooth variations

$$\delta\Phi(x,y,z,t_1) = 0, \quad \delta\Phi(x,y,z,t_2) = 0,$$
$$\delta Z(x,y,z,t_1) = 0, \quad \delta Z(x,y,z,t_2) = 0. \tag{2.23}$$

Proof is given in [151]. Its details could be useful for understanding the further derivations of the Miles-Lukovsky modal equation.

Specify the two unknowns as $\Phi = \Phi(x,y,z,t,\alpha_1)$ and $Z = Z(x,y,z,t,\alpha_2)$, where small parameters α_1 and α_2 are independent and their zeros correspond to an extremum of W. Substitution of Φ and Z into (2.22) yields a function of the two real variables, α_1 and α_2 ($W = W(\alpha_1, \alpha_2)$). The differential of $W(\alpha_1, \alpha_2)$ at $\alpha_1 = \alpha_2 = 0$ reads as

$$dW(0,0) = \left.\frac{\partial W}{\partial \alpha_1}\right|_{0,0} d\alpha_1 + \left.\frac{\partial W}{\partial \alpha_2}\right|_{0,0} d\alpha_2 = \delta W = \frac{\partial W}{\partial Z}\underbrace{\frac{\partial Z}{\partial \alpha_2}d\alpha_2}_{\delta Z} + \frac{\partial W}{\partial \Phi}\underbrace{\frac{\partial \Phi}{\partial \alpha_1}d\alpha_1}_{\delta \Phi}$$

$$= -\rho \int_{t_1}^{t_2} \left(-\int_{\Sigma(t)} \left[\frac{\partial \Phi}{\partial t} + \frac{1}{2}(\nabla\Phi)^2 - \nabla\Phi \cdot (\boldsymbol{v}_O + \boldsymbol{\omega} \times \boldsymbol{r}) + U_g \right] \frac{\partial Z/\partial t}{|\nabla Z|} dS \right.$$
$$\left. + \int_{Q(t)} \left[\nabla\Phi \cdot \nabla(\delta\Phi) + \frac{\partial(\delta\Phi)}{\partial t} - \nabla(\delta\Phi) \cdot (\boldsymbol{v}_O + \boldsymbol{\omega} \times \boldsymbol{r}) \right] dQ \right) dt = 0, \tag{2.24}$$

where we accounted for the gravity potential U_g and $(\boldsymbol{v}_O + \boldsymbol{\omega} \times \boldsymbol{r})$ being independent of α_1 and α_2.

Because δZ and $\delta\Phi$ are independent variables, one can choose, for a while, $\delta\Phi = 0$. By considering arbitrary $\delta Z \neq 0$, equality (2.24) deduces the dynamic boundary condition (2.12) on $\Sigma(t)$. Alternatively, when $\delta Z = 0$, $\delta\Phi \neq 0$, integrals in (2.24) can be modified by using Green's formula, Gauss' theorem and Reynolds' transport theorem,

$$\int_{Q(t)} \nabla\Phi \cdot \nabla(\delta\Phi) \, dQ = \int_{S(t)+\Sigma(t)} \delta\Phi \frac{\partial \Phi}{\partial n} \, dS - \int_{Q(t)} \nabla^2\Phi \delta\Phi \, dQ,$$

$$\int_{Q(t)} \boldsymbol{v}_O \cdot \nabla(\delta\Phi) \, dQ = \int_{Q(t)} \nabla(\boldsymbol{v}_O \cdot \boldsymbol{r}) \cdot \nabla(\delta\Phi) \, dQ = \int_{S(t)+\Sigma(t)} \delta\Phi(\boldsymbol{v}_O \cdot \boldsymbol{n}) \, dS,$$

$$\int_{Q(t)} (\boldsymbol{\omega} \times \boldsymbol{r}) \cdot \nabla(\delta\Phi) \, dQ = \int_{S(t)+\Sigma(t)} \delta\Phi((\boldsymbol{\omega} \times \boldsymbol{r}) \cdot \boldsymbol{n}) \, dS,$$

$$\int_{Q(t)} \frac{\partial(\delta\Phi)}{\partial t} \, dQ = \frac{d}{dt}\int_{Q(t)} \delta\Phi \, dQ + \int_{\Sigma(t)} (\delta\Phi)|_{\Sigma(t)} \frac{\partial Z/\partial t}{|\nabla Z|} \, dS,$$

respectively, to transform (2.24) to the variational equality

$$\delta W|_{\delta Z=0} = \rho \int_{t_1}^{t_2} \left(\int_{S(t)} \left[\frac{\partial \Phi}{\partial n} - (\boldsymbol{v}_O + \boldsymbol{\omega} \times \boldsymbol{r}) \cdot \boldsymbol{n} \right] \delta\Phi \, dS \right.$$
$$\left. - \int_{Q(t)} \nabla^2\Phi \delta\Phi \, dQ + \int_{\Sigma(t)} \left[\frac{\partial \Phi}{\partial n} - (\boldsymbol{v}_O + \boldsymbol{\omega} \times \boldsymbol{r}) \cdot \boldsymbol{n} + \frac{\partial Z}{\partial t}/|\nabla Z| \right] \delta\Phi \, dS \right) dt$$

Differential, variational, and modal statements 19

$$-\rho \int_{Q(t)} \delta\Phi \, dQ\Big|_{t=t_1}^{t=t_2} = 0, \quad (2.25)$$

where the term $\rho \int_{Q(t)} \delta\Phi \, dQ\big|_{t=t_1}^{t=t_2} = 0$ since $\delta\Phi = 0$ at $t = t_1, t_2$ according to (2.23). Variational equality (2.25) yields the kinematic boundary condition on $\Sigma(t) + S(t)$ and the Laplace equation in $Q(t)$. ∎

2.2.2 ROTATIONAL FLOWS

Theorem 2.1 can be generalised to rotational liquid flows by using the Clebsch potentials, $\varphi(x, y, z, t), m(x, y, z, t)$, and $\phi(x, y, z, t)$, which define the *absolute velocity field* $\boldsymbol{v} = (v_1, v_2, v_3, t)$ as follows

$$\boldsymbol{v} = \nabla\varphi + m\nabla\phi \quad (2.26)$$

(irrotational flows imply either $m = 0$ or $\phi = \text{const}$).

Remark 2.1. *Even though (2.26) does not give a unique representation of the velocity field (substitution $m := Cm, \phi := \phi/C$, where C is a non-zero constant, confirms that), the Clebsch potentials are assumed being three independent functions in the current section. Furthermore, as remarked in [62] (p. 47), the spatial derivatives in the introduced inertial (∂') and non-inertial (∂) coordinate systems remain the same, but the time-derivatives $(\partial'/\partial t'$ and $\partial/\partial t$, respectively) change, i.e.,*

$$\frac{\partial'}{\partial t} = \frac{\partial}{\partial t} - \boldsymbol{v}_b \cdot \nabla, \quad \frac{d'}{dt} = \frac{\partial'}{\partial t} + \boldsymbol{v} \cdot \nabla = \frac{\partial}{\partial t} + (\boldsymbol{v} - \boldsymbol{v}_b) \cdot \nabla, \quad (2.27)$$

where $\boldsymbol{v}_b$ is the rigid-body velocity by (2.6).

Henceforth, the Clebsch potentials and Z are smooth functions, which admit an analytical continuation through the smooth free surface $\Sigma(t)$ for any instant time t.

Based on relations (2.27) and [11] (p. 164), the following Lagrangian

$$L(\varphi, m, \phi, Z) = \int_{Q(t)} P \, dQ = -\rho \int_{Q(t)} \left[\frac{\partial'\varphi}{\partial t} + m\frac{\partial'\phi}{\partial t} + \frac{1}{2}|\boldsymbol{v}|^2 + U_g \right] dQ$$

$$= \rho \int_{Q(t)} \left[\frac{\partial\varphi}{\partial t} + m\frac{\partial\phi}{\partial t} - \boldsymbol{v}_b \cdot \boldsymbol{v} + \frac{1}{2}|\boldsymbol{v}|^2 + U_g \right] dQ \quad (2.28)$$

and the action

$$W(\varphi, m, \phi, Z) = \int_{t_1}^{t_2} \left[L - p_0 \int_{Q(t)} dQ \right] dt = \int_{t_1}^{t_2} \int_{Q(t)} (P - p_0) \, dQ \, dt \quad (2.29)$$

are introduced for any fixed instant time $t_1 < t < t_2$. The action (2.29) is defined on the Clebsch potentials and function Z. The Lagrange multiplier p_0 is a consequence of the volume conservation constraint (2.18).

Lemma 2.1

Under the assumption in Remark 2.1, the zero first variation
$$\delta_\varphi W = 0 \quad \text{subject to} \quad \delta\varphi|_{t=t_1,t_2} = 0 \qquad (2.30)$$
is equivalent to the kinematic relations consisting of the continuity equation
$$\nabla \cdot (\boldsymbol{v} - \boldsymbol{v}_b) \equiv \nabla \cdot \boldsymbol{v} = 0 \quad \text{in } Q(t) \qquad (2.31)$$
as well as the kinematic boundary conditions
$$(\boldsymbol{v} - \boldsymbol{v}_b) \cdot \boldsymbol{n} = 0 \quad \text{on} \quad S(t), \quad (\boldsymbol{v} - \boldsymbol{v}_b) \cdot \boldsymbol{n} = -\frac{\partial Z/\partial t}{|\nabla Z|} \quad \text{on } \Sigma(t) \qquad (2.32)$$
expressing that the normal velocity is defined by the rigid wall motins and the fluid particles remain on the free surface $\Sigma(t)$.

Proof. Deriving the first variation by φ is similar (but not the same) to that for the potential flows [62] (p. 58-59). Consequently using the Reynolds transport theorem, divergence theorems, and condition $\delta\varphi|_{t=t_1,t_2} = 0$ gives

$$\begin{aligned}
\delta_\varphi W = -\rho \int_{t_1}^{t_2} &\int_{Q(t)} \left(\frac{\partial(\delta\varphi)}{\partial t} + (\boldsymbol{v} - \boldsymbol{v}_b) \cdot \nabla(\delta\varphi) \right) dQ dt \\
&- \rho \int_{t_1}^{t_2} \left(\left[\frac{d}{dt} \int_{Q(t)} \delta\varphi \, dQ + \int_{\Sigma(t)} \frac{\partial Z/\partial t}{|\nabla Z|} \delta\varphi \, dS \right] \right. \\
&+ \left[\int_{S(t)+\Sigma(t)} (\boldsymbol{v} - \boldsymbol{v}_b) \cdot \boldsymbol{n} \, \delta\varphi \, dS - \int_{Q(t)} \nabla \cdot (\boldsymbol{v} - \boldsymbol{v}_b) \, \delta\varphi \, dQ \right] \right) dt \\
= -\rho \int_{t_1}^{t_2} &\left(\int_{\Sigma(t)} \left[(\boldsymbol{v} - \boldsymbol{v}_b) \cdot \boldsymbol{n} + \frac{\partial Z/\partial t}{|\nabla Z|} \right] \delta\varphi \, dS \right. \\
&\left. + \int_{S(t)} (\boldsymbol{v} - \boldsymbol{v}_b) \cdot \boldsymbol{n} \, \delta\varphi \, dS - \int_{Q(t)} \nabla \cdot (\boldsymbol{v} - \boldsymbol{v}_b) \, \delta\varphi \, dQ \right) = 0, \qquad (2.33)
\end{aligned}$$

which deduces (2.31) and (2.32) by using the standart calculus of variables. ∎

Lemma 2.2

Under the assumption in Remark 2.1, the zero first variation
$$\delta_m W = 0 \qquad (2.34)$$
is equivalent to the equation
$$\frac{d'\phi}{dt} \equiv \frac{\partial'\phi}{\partial t} + \boldsymbol{v} \cdot \nabla\phi \equiv \frac{\partial\phi}{\partial t} + (\boldsymbol{v} - \boldsymbol{v}_b) \cdot \nabla\phi = 0 \quad \text{in } Q(t), \qquad (2.35)$$

which says that the Clebsch potential ϕ remains constant during the trajectories of liquid particles (the vortex line moves with the liquid and always contains the same particles).

Proof. The variation by m derives the variational equality

$$\delta_m W = -\rho \int_{t_1}^{t_2} \int_{Q(t)} \left[\frac{\partial \phi}{\partial t} + (\boldsymbol{v} - \boldsymbol{v}_b) \cdot \nabla \phi \right] \delta m \, \mathrm{d}Q \, \mathrm{d}t = 0, \qquad (2.36)$$

which proves the lemma. ∎

Lemma 2.3

Under the assumption in Remark 2.1, the zero first variation

$$\delta_\phi W = 0 \quad \text{subject to} \quad \delta\phi|_{t_1, t_2} = 0 \qquad (2.37)$$

and the kinematic problem (2.31), (2.32) is equivalent to

$$\frac{\mathrm{d}'m}{\mathrm{d}t} \equiv \frac{\partial' m}{\partial t} + \boldsymbol{v} \cdot \nabla m \equiv \frac{\partial m}{\partial t} + (\boldsymbol{v} - \boldsymbol{v}_b) \cdot \nabla m = 0 \quad \text{in } Q(t), \qquad (2.38)$$

which has the same meaning that (2.35) has but for the Clebsch potential m.

Proof. The first variation by ϕ reads as

$$\delta_\phi W = -\rho \int_{t_1}^{t_2} \int_{Q(t)} m \left(\frac{\partial(\delta\phi)}{\partial t} + (\boldsymbol{v} - \boldsymbol{v}_b) \cdot \nabla(\delta\phi) \right) \mathrm{d}Q \, \mathrm{d}t$$

$$= -\rho \int_{t_1}^{t_2} \left(\left[\frac{\mathrm{d}}{\mathrm{d}t} \int_{Q(t)} m \delta\phi \, \mathrm{d}Q - \int_{Q(t)} \frac{\partial m}{\partial t} \delta\phi \, \mathrm{d}Q + \int_{\Sigma(t)} \frac{\partial Z/\partial t}{|\nabla Z|} m \, \delta\phi \, \mathrm{d}S \right] \right.$$

$$+ \left[\int_{S(t)+\Sigma(t)} m(\boldsymbol{v} - \boldsymbol{v}_b) \cdot \boldsymbol{n} \, \delta\phi \, \mathrm{d}S \right]$$

$$\left. - \int_{Q(t)} \delta\phi \left(m \nabla \cdot (\boldsymbol{v} - \boldsymbol{v}_b) + (\boldsymbol{v} - \boldsymbol{v}_b) \cdot \nabla m \right) \mathrm{d}Q \right) \mathrm{d}t$$

$$= \rho \int_{t_1}^{t_2} \int_{Q(t)} \delta\phi \left[\frac{\partial m}{\partial t} + (\boldsymbol{v} - \boldsymbol{v}_b) \cdot \nabla m \right] \mathrm{d}Q \, \mathrm{d}t = 0, \quad (2.39)$$

where the Reynolds transport and divergence theorems, the zero variation condition (2.37) at $t = t_1$ and t_2 as well as the kinematic conditions (2.31) and (2.32) were used. The last line of variational equality (2.39) proves the lemma. ∎

Remark 2.2. *In contrast to the Bateman-Luke formulation for irrotational flows by Theorem 2.1, the function P adopted in the Lagrangian (2.28) is, generally speaking, not the pressure and, moreover, cannot be treated as the* pressure for arbitrary Clebsch's potentials. One can show that the pressure $p = P + f(t)$ ($f(t)$ is an arbitrary function) when (2.35) and (2.38) are satisfied. In other words, when assuming (2.35) and (2.38), the Euler equation

$$\frac{d'\boldsymbol{v}}{dt} = -\frac{1}{\rho}(\nabla P + \nabla U) \quad \text{in} \quad Q(t) \tag{2.40}$$

is formally fulfilled. This fact follows from the left-hand side of (2.40) by

$$\frac{d'(\nabla\varphi + m\nabla\phi)}{dt} = \left[\nabla\frac{\partial'\varphi}{\partial t} + m\nabla\frac{\partial'\phi}{\partial t} + \frac{\partial' m}{\partial t}\nabla\phi\right] + \underbrace{\boldsymbol{v}\cdot\nabla(\nabla\varphi + m\nabla\phi)}_{\boldsymbol{v}\cdot\nabla\nabla\varphi + m\boldsymbol{v}\cdot\nabla\nabla\phi + \nabla\phi(\nabla m\cdot\boldsymbol{v})}$$

$$= \underline{\nabla\frac{\partial'\varphi}{\partial t} + m\nabla\frac{\partial'\phi}{\partial t} + \boldsymbol{v}\cdot\nabla\nabla\varphi + m\boldsymbol{v}\cdot\nabla\nabla\phi + \nabla\phi(\nabla m\cdot\boldsymbol{v})} + \nabla\phi\frac{d'm}{dt}$$

and the right-hand side (after annihilating the U-term)

$$\nabla\left(\frac{\partial'\varphi}{\partial t} + m\frac{\partial'\phi}{\partial t} + \frac{1}{2}|\boldsymbol{v}|^2\right) = \left[\nabla\frac{\partial'\varphi}{\partial t} + m\nabla\frac{\partial'\phi}{\partial t} + \frac{\partial'\phi}{\partial t}\nabla m\right]$$
$$+ \boldsymbol{v}\cdot\nabla\nabla\varphi + m\boldsymbol{v}\cdot\nabla\nabla\phi + \nabla m(\nabla\phi\cdot\boldsymbol{v})$$

$$= \underline{\nabla\frac{\partial'\varphi}{\partial t} + m\nabla\frac{\partial'\phi}{\partial t} + \boldsymbol{v}\cdot\nabla\nabla\varphi + m\boldsymbol{v}\cdot\nabla\nabla\phi + \nabla\phi(\nabla m\cdot\boldsymbol{v})} + \nabla m\frac{d'\phi}{dt},$$

in which the underlined terms are identical, but the residual terms vanish as (2.35) and (2.38) hold true.

Theorem 2.2

Under the assumption in Remark 2.1, the zero first variation of the action (2.29)

$$\delta W = \delta_\varphi W + \delta_m W + \delta_\phi W + \delta_Z W \tag{2.41}$$

subject to

$$\delta\varphi|_{t_1,t_2} = \delta\phi_{t_1,t_2} = 0 \tag{2.42}$$

is equivalent to the sloshing boundary value problem for an incompressible ideal liquid with rotational flows, which includes the kinematic relations (2.31) and (2.32), two equations (2.35) and (2.38) expressing the fact that the Clebsch potentials ϕ and m are constant along the vortex lines, as well as the dynamic boundary condition

$$p - p_0 = -\rho\left(\frac{\partial'\varphi}{\partial t} + m\frac{\partial'\phi}{\partial t} - \boldsymbol{v}_b\cdot\boldsymbol{v} + \frac{1}{2}|\boldsymbol{v}|^2 + U_g\right) - p_0 \quad \text{on} \quad \Sigma(t) \tag{2.43}$$

establishing that the pressure equals the ullage pressure p_0 on the free surface. The volume conservation condition (2.18) should be added to the sloshing problem.

Differential, variational, and modal statements 23

Proof. The theorem follows from Lemmas 2.1–2.3, Remark 2.2 establishing that P in the Lagrangian can be treated as the pressure p, and the variational equality

$$\delta_Z W = -\int_{t_1}^{t_2}\int_{\Sigma(t)}(p-p_0)\frac{\delta Z}{|\nabla Z|}\,\mathrm{d}Q\,\mathrm{d}t = 0, \qquad (2.44)$$

which derives the dynamic boundary condition (2.43). ∎

2.3 MILES-LUKOVSKY MODAL EQUATIONS

We can use the Bateman-Luke variational principle by Theorem 2.1 to get the so-called modal system (a system of ordinary differential equations with respect to the hydrodynamic generalised coordinates and velocities). The system is fully equivalent to the original free-boundary problem. The derivations were reported in [143], [151], and [54].

2.3.1 MODAL REPRESENTATION OF THE SOLUTION

Assume the free surface $\Sigma(t)$ is presented by the single-valued equation $z = \zeta(x,y,t)$ $(Z(x,y,z,t) = z - \zeta(x,y,t) = 0)$ in the non-inertial coordinate system $Oxyz$ rigidly fixed with the tank as follows

$$\zeta(x,y,t) = \sum_{i=1}^{\infty}\beta_i(t)f_i(x,y). \qquad (2.45)$$

Here, the functional basis $\{f_i(x,y)\}$ is not necessarily the natural sloshing modes (see details in Chapter 3). The time-dependent functions $\{\beta_i\}$ play the role of the hydrodynamic *generalised coordinates*. The necessary condition for $\{f_i(x,y)\}$ consists of preserving the liquid volume (2.18), which means for (2.45) that

$$\int_{\Sigma_0}f_i(x,y)\,\mathrm{d}x\mathrm{d}y = 0 \ \ \text{for any } i,$$

where Σ_0 is the mean free surface. This condition is equivalent to (2.18).

A suitable functional (Fourier–type) representation is also required for the velocity potential $\Phi(x,y,z,t)$. Constructing it should account for that Φ is not zero for the 'frozen' free surface; the corresponding solution reflects then translational motions of fluid particles plus the Joukowski solution associated with the Stokes-Joukowski potentials, i.e.,

$$\Phi(x,y,z,t) = \boldsymbol{v}_O(t)\cdot\boldsymbol{r} + \boldsymbol{\omega}(t)\cdot\boldsymbol{\Omega}(x,y,z,\{\beta_i(t)\}) + \underbrace{\sum_{n=1}^{\infty}R_n(t)\varphi_n(x,y,z)}_{\varphi(x,y,z,t)}, \qquad (2.46)$$

where the first term implies the translational motions of liquid particles, φ is caused by the free-surface motions, and the vector-function

$$\boldsymbol{\Omega}(x,y,z,\{\beta_i(t)\}) = \Omega_1(x,y,z,\{\beta_i(t)\})\boldsymbol{e}_1(t) + \Omega_2(x,y,z,\{\beta_i(t)\})\boldsymbol{e}_2(t)$$
$$+ \Omega(x,y,z,\{\beta_i(t)\})\boldsymbol{e}_3(t) \quad (2.47)$$

($\boldsymbol{e}_i$, $i=1,2,3$ are the $Oxyz$-coordinate units) introduces the generalised Stokes-Joukowski potentials $\Omega_i(x,y,z,\{\beta_i(t)\})$, which are functions of spatial coordinates and the hydrodynamic generalised coordinates because the potentials follow from the following Neumann boundary value problem in the instantly-frozen liquid domain,

$$\nabla^2 \boldsymbol{\Omega} = 0 \text{ in } Q(t), \quad \frac{\partial \boldsymbol{\Omega}}{\partial n} = \boldsymbol{r} \times \boldsymbol{n} \text{ on } S(t) + \Sigma(t) \Rightarrow$$
$$\frac{\partial \Omega_1}{\partial n} = yn_z - zn_y; \quad \frac{\partial \Omega_2}{\partial n} = zn_x - xn_z; \quad \frac{\partial \Omega_3}{\partial n} = xn_y - yn_x \text{ on } S(t) + \Sigma(t), \quad (2.48)$$

where $\boldsymbol{n} = (n_x, n_y, n_z)$ is the outer normal unit vector and $\boldsymbol{r} = (x,y,z)$.

The functional basis $\{\varphi_n(x,y,z)\}$ in the modal representation (2.46) must be complete for any admissible liquid shape $Q(t)$. When it comes to practical applications, $\{\varphi_n(x,y,z)\}$ usually coincides with the natural sloshing modes. Then Φ by (2.46) automatically satisfies the Laplace equation and the boundary conditions on the wetted tank surface. But the kinematic and dynamic boundary conditions on the free surface are not satisfied.

Because the kinematic and dynamic boundary conditions naturally follow from the Bateman-Luke variational principle, it can be adopted as a mathematical tool for deriving the modal equations, which couple the hydrodynamic generalised coordinates $\beta_i(t)$ and velocities $R_n(t)$. These differential equations, at least, their dynamic component, can be interpreted as the Euler-Lagrange equation of the second kind (with respect to β_i and R_n) for the considered hydromechanical system.

2.3.2 EULER-LAGRANGE EQUATION

After substitution (2.46) into (2.22), the Lagrangian takes the following form

$$L = -\rho \int_{Q(t)} \left[\frac{d\boldsymbol{v}_O}{dt} \cdot \boldsymbol{r} + \frac{\partial}{\partial t}(\boldsymbol{\omega} \cdot \boldsymbol{\Omega}) \right.$$
$$+ \frac{1}{2}\nabla(\boldsymbol{\omega} \cdot \boldsymbol{\Omega}) \cdot \nabla(\boldsymbol{\omega} \cdot \boldsymbol{\Omega}) - \boldsymbol{\omega} \cdot (\boldsymbol{r} \times \nabla(\boldsymbol{\omega} \cdot \boldsymbol{\Omega})) - \frac{1}{2}v_O^2 - \boldsymbol{\omega} \cdot (\boldsymbol{r} \times \boldsymbol{v}_O)$$
$$\left. - \boldsymbol{\omega} \cdot (\boldsymbol{r} \times \nabla\varphi) + \nabla(\boldsymbol{\omega} \cdot \boldsymbol{\Omega}) \cdot \varphi \right] dQ + L_r, \quad (2.49)$$

where

$$L_r = -\rho \int_{Q(t)} \left[\frac{\partial \varphi}{\partial t} + \frac{1}{2}(\nabla\varphi)^2 + U_g \right] dQ. \quad (2.50)$$

Integral expressions in (2.49) require some simplifications based on the vector algebra and the Gauss theorem. The two terms in the integrand of

(2.49) cancel each other, that is,

$$\int_{Q(t)} [-(\boldsymbol{\omega} \times \boldsymbol{r}) \cdot \nabla\varphi + \nabla(\boldsymbol{\omega} \cdot \boldsymbol{\Omega}) \cdot \nabla\varphi] dQ$$
$$= \int_{S(t)+\Sigma(t)} \left(\frac{\partial(\boldsymbol{\omega} \cdot \boldsymbol{\Omega})}{\partial n} - (\boldsymbol{\omega} \times \boldsymbol{r}) \cdot \boldsymbol{n} \right) \varphi dS = 0. \quad (2.51)$$

Other components in (2.49) can be simplified to a more compact form after introducing the inertia tensor $J^1 = J^1(x, y, z, \{\beta_i(t)\})$, whose components are

$$J_{11}^1 = \rho \int_{Q(t)} \left(y \frac{\partial \Omega_1}{\partial z} - z \frac{\partial \Omega_1}{\partial y} \right) dQ = \rho \int_{S(t)+\Sigma(t)} \Omega_1 \frac{\partial \Omega_1}{\partial n} dS,$$

$$J_{22}^1 = \rho \int_{Q(t)} \left(z \frac{\partial \Omega_2}{\partial x} - x \frac{\partial \Omega_2}{\partial z} \right) dQ = \rho \int_{S(t)+\Sigma(t)} \Omega_2 \frac{\partial \Omega_2}{\partial n} dS,$$

$$J_{33}^1 = \rho \int_{Q(t)} \left(x \frac{\partial \Omega_3}{\partial y} - y \frac{\partial \Omega_3}{\partial x} \right) dQ = \rho \int_{S(t)+\Sigma(t)} \Omega_3 \frac{\partial \Omega_3}{\partial n} dS,$$

$$J_{12}^1 = J_{21}^1 = \rho \int_{Q(t)} \left(z \frac{\partial \Omega_1}{\partial x} - x \frac{\partial \Omega_1}{\partial z} \right) dQ = \rho \int_{Q(t)} \left(y \frac{\partial \Omega_2}{\partial z} - z \frac{\partial \Omega_2}{\partial y} \right) dQ$$
$$= \rho \int_{S(t)+\Sigma(t)} \Omega_1 \frac{\partial \Omega_2}{\partial n} dS = \rho \int_{S(t)+\Sigma(t)} \Omega_2 \frac{\partial \Omega_1}{\partial n} dS,$$

$$J_{13}^1 = J_{31}^1 = \rho \int_{Q(t)} \left(x \frac{\partial \Omega_1}{\partial y} - y \frac{\partial \Omega_1}{\partial x} \right) dQ = \rho \int_{Q(t)} \left(y \frac{\partial \Omega_3}{\partial z} - z \frac{\partial \Omega_3}{\partial y} \right) dQ$$
$$= \rho \int_{S(t)+\Sigma(t)} \Omega_1 \frac{\partial \Omega_3}{\partial n} dS = \rho \int_{S(t)+\Sigma(t)} \Omega_3 \frac{\partial \Omega_1}{\partial n} dS,$$

$$J_{23}^1 = J_{32}^1 = \rho \int_{Q(t)} \left(x \frac{\partial \Omega_2}{\partial y} - y \frac{\partial \Omega_2}{\partial x} \right) dQ = \rho \int_{Q(t)} \left(z \frac{\partial \Omega_3}{\partial x} - x \frac{\partial \Omega_3}{\partial z} \right) dQ$$
$$= \rho \int_{S(t)+\Sigma(t)} \Omega_2 \frac{\partial \Omega_3}{\partial n} dS = \rho \int_{S(t)+\Sigma(t)} \Omega_3 \frac{\partial \Omega_2}{\partial n} dS, \quad (2.52)$$

where the first Green identity and the Gauss theorem were used. The inertia tensor is associated with the following quadratic form

$$-\frac{1}{2}\omega_1^2 J_{11}^1 - \frac{1}{2}\omega_2^2 J_{22}^1 - \frac{1}{2}\omega_3^2 J_{33}^1 - \omega_1\omega_2 J_{12}^1 - \omega_1\omega_3 J_{13}^1 - \omega_2\omega_3 J_{23}^1$$
$$+ \frac{1}{2}\rho \int_{S(t)+\Sigma(t)} (\boldsymbol{\omega} \cdot \boldsymbol{\Omega}) \left(\frac{\partial \boldsymbol{\Omega}}{\partial n} \cdot \boldsymbol{\omega} \right) dS =$$
$$= \frac{1}{2}\rho \int_{Q(t)} \left(\frac{1}{2} \nabla(\boldsymbol{\omega} \cdot \boldsymbol{\Omega}) \cdot \nabla(\boldsymbol{\omega} \cdot \boldsymbol{\Omega}) - \boldsymbol{\omega} \cdot (\boldsymbol{r} \times \nabla(\boldsymbol{\omega} \cdot \boldsymbol{\Omega})) \right) dQ. \quad (2.53)$$

After introducing the inertia tensor, (2.49) possesses the form

$$L = -[\dot{v}_{O1}l_1 + \dot{v}_{O2}l_2 + \dot{v}_{O3}l_3 + \dot{\omega}_1 l_{1\omega} + \dot{\omega}_2 l_{2\omega} + \dot{\omega}_3 l_{3\omega} + \omega_1 l_{1\omega t} + \omega_2 l_{2\omega t} + \omega_3 l_{3\omega t}$$
$$- \frac{1}{2}(\omega_1^2 J_{11}^1 + \omega_2^2 J_{22}^1 + \omega_3^2 J_{33}^1) - \omega_1 \omega_2 J_{12}^1 - \omega_1 \omega_3 J_{13}^1 - \omega_2 \omega_3 J_{23}^1$$
$$- \frac{1}{2} M_l (v_{O1}^2 + v_{O2}^2 + v_{O3}^2) + (\omega_2 v_{O3} - \omega_3 v_{O2}) l_1$$
$$+ (\omega_3 v_{O1} - \omega_1 v_{O3}) l_2 + (\omega_1 v_{O2} - \omega_2 v_{O1}) l_3] + L_r, \quad (2.54)$$

where M_l is the liquid mass and

$$l_{k\omega} = \rho \int_{Q_t} \Omega_k dQ, \quad l_{k\omega t} = \rho \int_{Q_t} \frac{\partial \Omega_k}{\partial t} dQ,$$
$$l_1 = \rho \int_{Q(t)} x dQ, \quad l_2 = \rho \int_{Q(t)} y dQ, \quad l_3 = \rho \int_{Q(t)} z dQ. \quad (2.55)$$

The vectors $\boldsymbol{l} = (l_1, l_2, l_3), \boldsymbol{l_\omega} = (l_{1\omega}, l_{2\omega}, l_{3\omega})$, and $\boldsymbol{l_{\omega t}} = (l_{1\omega t}, l_{2\omega t}, l_{3\omega t})$ are functions of $\{\beta_i\}$ and $\{\dot{\beta}_i\}$. Moreover, component (2.50) should account for the modal representation (2.46), from where we can get

$$L_r = -\rho \int_{Q(t)} \left[\sum_{n=1}^{\infty} \dot{R}_n \varphi_n + \frac{1}{2} \sum_{n,k=1}^{\infty} R_n R_k (\nabla \varphi_n \cdot \nabla \varphi_k) + U_g \right] dQ$$
$$= -\left[\sum_{n=1}^{\infty} D_n \dot{R}_n + \frac{1}{2} \sum_{n,k=1}^{\infty} D_{nk} R_n R_k - g_1 l_1 - g_2 l_2 - g_3 l_3 - m_l g \cdot r_0' \right], \quad (2.56)$$

where

$$A_n = \rho \int_{Q(t)} \varphi_n dQ, \quad A_{nk} = A_{kn} = \rho \int_{Q(t)} (\nabla \varphi_n \cdot \nabla \varphi_k) dQ. \quad (2.57)$$

The Bateman-Luke variational principle (Theorem 2.1), equations (2.22) and (2.23) treat L as a function of two *independent* variables $\zeta(x, y, t)$ and $\Phi(x, y, z, t)$. After substitution of (2.45) and (2.46), L is expressed by equations (2.54) and (2.56) and becomes a function of the generalised coordinates $\{\beta_i\}$ and velocities $\{R_n\}$. Note that integrals in (2.52), (2.55), and (2.57) are functions of $\{\beta_i\}$. Independent variations of the action $W = \int_{t_1}^{t_2} L dt$ (through $\delta \zeta$ and $\delta \Phi$) must be related to independent variations by $\delta \beta_i$ and δR_n, respectively. This gives the following variational equality

$$\delta W = \int_{t_1}^{t_2} \left[\sum_n A_n \delta \dot{R}_n + \sum_n A_{nk} R_k \delta R_n \right.$$
$$+ \left(\sum_n \dot{R}_n \frac{\partial A_n}{\partial \beta_i} + \omega_1 \frac{\partial l_{1\omega t}}{\partial \beta_i} + \omega_2 \frac{\partial l_{2\omega t}}{\partial \beta_i} + \omega_3 \frac{\partial l_{3\omega t}}{\partial \beta_i} \right.$$

$$+\frac{1}{2}\sum_{n,k}R_nR_k\frac{\partial A_{nk}}{\partial \beta_i}+\dot{\omega}_1\frac{\partial l_{1\omega}}{\partial \beta_i}+\dot{\omega}_2\frac{\partial l_{2\omega}}{\partial \beta_i}+\dot{\omega}_3\frac{\partial l_{3\omega}}{\partial \beta_i}$$

$$+(\dot{v}_{O1}-g_1+\omega_2 v_{O3}-\omega_3 v_{O2})\frac{\partial l_1}{\partial \beta_i}+(\dot{v}_{O2}-g_2+\omega_3 v_{O1}-\omega_1 v_{O3})\frac{\partial l_2}{\partial \beta_i}$$

$$+(\dot{v}_{O3}-g_3+\omega_1 v_{O2}-\omega_2 v_{O1})\frac{\partial l_3}{\partial \beta_i}-\frac{1}{2}\omega_1^2\frac{\partial J_{11}^1}{\partial \beta_i}-\frac{1}{2}\omega_2^2\frac{\partial J_{22}^1}{\partial \beta_i}-\frac{1}{2}\omega_3^2\frac{\partial J_{33}^1}{\partial \beta_i}$$

$$-\omega_1\omega_2\frac{\partial J_{12}^1}{\partial \beta_i}-\omega_1\omega_3\frac{\partial J_{13}^1}{\partial \beta_i}-\omega_2\omega_3\frac{\partial J_{23}^1}{\partial \beta_i}\bigg)\delta\beta_i$$

$$+\left(\omega_1\frac{\partial l_{1\omega t}}{\partial \dot{\beta}_i}+\omega_2\frac{\partial l_{2\omega t}}{\partial \dot{\beta}_i}+\omega_3\frac{\partial l_{3\omega t}}{\partial \dot{\beta}_i}\right)\delta\dot{\beta}_i\bigg]dt=0,\ i\geq 1. \quad (2.58)$$

Quantities, which are proportional to $\delta\dot{R}_n$ and $\delta\dot{\beta}_i$ in the last expression of (2.58), can be integrated by parts, and by using the boundary conditions

$$\delta R_n(t_1)=\delta R_n(t_2)=\delta\beta_i(t_1)=\delta\beta_i(t_2)=0,$$

converted to expressions through δR_n and $\delta\beta_i$ instead of $\delta\dot{R}_n$ and $\delta\dot{\beta}_i$. Utilizing the standard way, we can then derive the following infinite system of nonlinear ordinary differential equations (*Miles-Lukovsky*'s equations) with respect to $\{R_n\}$ and $\{\beta_i\}$:

$$\sum_i\frac{\partial A_n}{\partial \beta_i}\dot{\beta}_i-\sum_k R_k A_{nk}=0,\ n=1,2,\ldots, \quad (2.59)$$

$$\sum_n \dot{R}_n\frac{\partial A_n}{\partial \beta_i}+\frac{1}{2}\sum_n\sum_k\frac{\partial A_{nk}}{\partial \beta_i}R_nR_k+\dot{\omega}_1\frac{\partial l_{1\omega}}{\partial \beta_i}+\dot{\omega}_2\frac{\partial l_{2\omega}}{\partial \beta_i}+\dot{\omega}_3\frac{\partial l_{3\omega}}{\partial \beta_i}$$

$$+\omega_1\frac{\partial l_{1\omega t}}{\partial \beta_i}+\omega_2\frac{\partial l_{2\omega t}}{\partial \beta_i}+\omega_3\frac{\partial l_{3\omega t}}{\partial \beta_i}-\frac{d}{dt}\left(\omega_1\frac{\partial l_{1\omega t}}{\partial \dot{\beta}_i}+\omega_2\frac{\partial l_{2\omega t}}{\partial \dot{\beta}_i}+\omega_3\frac{\partial l_{3\omega t}}{\partial \dot{\beta}_i}\right)$$

$$+(\dot{v}_{O1}-g_1+\omega_2 v_{O3}-\omega_3 v_{O2})\frac{\partial l_1}{\partial \beta_i}+(\dot{v}_{O2}-g_2+\omega_3 v_{O1}-\omega_1 v_{O3})\frac{\partial l_2}{\partial \beta_i}$$

$$+(\dot{v}_{O3}-g_3+\omega_1 v_{O2}-\omega_2 v_{O1})\frac{\partial l_3}{\partial \beta_i}-\frac{1}{2}\omega_1^2\frac{\partial J_{11}^1}{\partial \beta_i}-\frac{1}{2}\omega_2^2\frac{\partial J_{22}^1}{\partial \beta_i}-\frac{1}{2}\omega_3^2\frac{\partial J_{33}^1}{\partial \beta_i}$$

$$-\omega_1\omega_2\frac{\partial J_{12}^1}{\partial \beta_i}-\omega_1\omega_3\frac{\partial J_{13}^1}{\partial \beta_i}-\omega_2\omega_3\frac{\partial J_{23}^1}{\partial \beta_i}=0,\ i=1,2,\ldots. \quad (2.60)$$

If the tank has an upright cylindrical shape in the vicinity of the mean free surface Σ_0, the value of $\partial l_k/\partial \delta_i$ can be written as

$$\frac{\partial l_3}{\partial \beta_i}=\rho\int_{\Sigma_0}f_i^2 dS\,\beta_i=\lambda_{3i}\beta_i,\quad \frac{\partial l_2}{\partial \beta_i}=\rho\int_{\Sigma_0}yf_i^2 dS=\lambda_{2i},$$

$$\frac{\partial l_1}{\partial \beta_i}=\rho\int_{\Sigma_0}xf_i^2 dS=\lambda_{1i}. \quad (2.61)$$

For translational tank motions, the system (2.59)–(2.60) was independently derived by Miles and Lukovsky in 1976. Its final form was obtained by Lukovsky. An alternative derivation was proposed in [54].

2.3.3 LUKOVSKY'S FORMULAS FOR THE HYDRODYNAMIC FORCE AND MOMENT

Following [62], one can derive famous Lukovsky's formulas for the resulting hydrodynamic force

$$F(t) = \int_{S(t)+\Sigma(t)} p\boldsymbol{n} \, dS \qquad (2.62)$$

and moment (relative to the origin O)

$$\boldsymbol{M}_O(t) = \int_{Q(t)} \boldsymbol{r} \times (p\boldsymbol{n}) \, dS, \qquad (2.63)$$

which are caused by the hydrodynamic loads on the wetted tank surface.

Remark 2.3. *Dealing with Lukovsky's formulas needs the two different time derivatives in inertial and non-inertial coordinate systems, which are marked by the dot and the star, respectively. By definition, if $\boldsymbol{B}(t) = B_1(t)\boldsymbol{e}_1(t) + B_2(t)\boldsymbol{e}_2(t) + B_3(t)\boldsymbol{e}_3(t)$ is a time-depending vector, and $\boldsymbol{e}_i(t)$ are the unit vectors of the non-inertial coordinate system,*

$$\dot{\boldsymbol{B}}(t) = \overset{*}{\boldsymbol{B}}(t) + \boldsymbol{\omega}(t) \times \boldsymbol{B}(t). \qquad (2.64)$$

The derivation of the first Lukovsky formula requires the hydrodynamic momentum,

$$\boldsymbol{M}(t) = \rho \int_{Q(t)} \boldsymbol{v} \, dQ = \rho \int_{Q'(t)} \boldsymbol{v} \, dQ' \qquad (2.65)$$

and Eq. (2.57) in [62],

$$\boldsymbol{F} = -\dot{\boldsymbol{M}} + M_l \boldsymbol{g}, \qquad (2.66)$$

which links the hydrodynamic momentum (2.65) and the hydrodynamic force (2.62).

Expression (2.65) should be rewritten in the non-inertial coordinate system $Oxyz$, where $\boldsymbol{v} = \boldsymbol{v}_O + \boldsymbol{\omega} \times \boldsymbol{r} + \boldsymbol{v}_r$ is the absolute velocity and $\boldsymbol{v}_r$ is the relative velocity field, that is,

$$\boldsymbol{M} = M_l \boldsymbol{v}_O + \rho \boldsymbol{\omega} \times \int_{Q(t)} \boldsymbol{r} \, dQ + \rho \int_{Q(t)} \boldsymbol{v}_r \, dQ. \qquad (2.67)$$

The last integral of (2.67) transforms to

$$\rho \int_{Q(t)} \boldsymbol{v}_r \, dQ = \rho \int_{\Sigma(t)} \boldsymbol{r}(\boldsymbol{v}_r \cdot \boldsymbol{n}) \, dS + \rho \int_{S(t)} \boldsymbol{r} \underbrace{(\boldsymbol{v}_r \cdot \boldsymbol{n})}_{0} \, dS = \rho \frac{d^*}{dt} \int_{Q(t)} \boldsymbol{r} \, dQ; \qquad (2.68)$$

here, we utilised the equality $\nabla \cdot (\boldsymbol{v}_r \boldsymbol{n}) = \boldsymbol{v}_r$ for the incompressible liquid, so that the normal relative velocity on the tank surface is zero, $\boldsymbol{v}_r \cdot \boldsymbol{n} = 0$ on $S(t)$ as well as the Gauss theorem and the Reynolds transport theorem.

Expressions (2.68) and (2.67) contain the volume integral over $\boldsymbol{r}$. When introducing the vector

$$\boldsymbol{r}_{lC}(t) = \left(\boldsymbol{e}_1 \int_{Q(t)} x \mathrm{d}Q + \boldsymbol{e}_2 \int_{Q(t)} y \mathrm{d}Q + \boldsymbol{e}_3 \int_{Q(t)} z \mathrm{d}Q \right) \Big/ \underbrace{\int_{Q(t)} \mathrm{d}Q}_{V_l}, \quad (2.69)$$

which gives the instant position of the *liquid mass centre* in the body-fixed coordinate system, (2.67) can be rewritten in the form

$$\boldsymbol{M} = M_l \left(\boldsymbol{v}_O + \boldsymbol{\omega} \times \boldsymbol{r}_{lC} + \overset{*}{\boldsymbol{r}}_{lC} \right). \quad (2.70)$$

Remembering the time-differentiation rule (2.64) and the fact that $\dot{\boldsymbol{\omega}} = \overset{*}{\boldsymbol{\omega}}$, substituting (2.70) into (2.66) and gathering similar terms derive

$$\boldsymbol{F}(t) = M_l \boldsymbol{g} - M_l \left[\overset{*}{\boldsymbol{v}}_O + \boldsymbol{\omega} \times \boldsymbol{v}_O + \boldsymbol{\omega} \times (\boldsymbol{\omega} \times \boldsymbol{r}_{lC}) + \dot{\boldsymbol{\omega}} \times \boldsymbol{r}_{lC} \right.$$
$$\left. + 2\boldsymbol{\omega} \times \overset{*}{\boldsymbol{r}}_{lC} + \overset{**}{\boldsymbol{r}}_{lC} \right], \quad (2.71)$$

where $\boldsymbol{\omega} \times (\boldsymbol{\omega} \times \boldsymbol{r}_{lC})$ is the centripetal acceleration and $2\boldsymbol{\omega} \times \overset{*}{\boldsymbol{r}}_{lC}$ is the Coriolis acceleration. Expression (2.71) represents the famous Lukovsky formula for the hydrodynamic force, which was derived by the author only for potential flows of an incompressible liquid, but the book [62] proved it for an arbitrary incompressible ideal liquid. The formula (2.71) should be valid for sloshing accompanied by overturning waves, free-surface fragmentation and other very special phenomena.

When the closed tank is completely filled with liquid, $Q(t)$ coincides with the whole tank cavity and does not depend on the time, therefore,

$$\boldsymbol{r}_{lC} = \boldsymbol{r}_{lC_0} = \mathrm{const}, \ \overset{*}{\boldsymbol{r}}_{lC} = 0,$$

(the liquid mass centre is motionless relative to the tank and coincides with the the hydrostatic liquid mass centre $\boldsymbol{r}_{lC_0}$). This simplifies (2.71) to the form

$$\boldsymbol{F}^{filled}(t) = M_l \boldsymbol{g} - M_l \left[\overset{*}{\boldsymbol{v}}_O + \boldsymbol{\omega} \times \boldsymbol{v}_O + \boldsymbol{\omega} \times (\boldsymbol{\omega} \times \boldsymbol{r}_{lC_0}) + \overset{*}{\boldsymbol{\omega}} \times \boldsymbol{r}_{lC_0} \right]. \quad (2.72)$$

The book [62] also presents derivation of the Lukovsky formula for the hydrodynamic moment relative to the origin O (definition (2.63)) by using analytical relations for the angular momentum

$$\boldsymbol{G}_O(t) = \rho \int_{Q(t)} \boldsymbol{r} \times \boldsymbol{v} \, \mathrm{d}Q = \rho \int_{Q'(t)} (\boldsymbol{r}' - \boldsymbol{r}_O) \times \boldsymbol{v} \, \mathrm{d}Q'. \quad (2.73)$$

Consequent time derivation and usage of the Reynolds transportation theorem yield

$$\dot{\boldsymbol{G}}_O = \overset{*}{\boldsymbol{G}}_O + \boldsymbol{\omega} \times \boldsymbol{G}_O = \rho \int_{Q'(t)} (\boldsymbol{r}' - \boldsymbol{r}_O(t)) \times \left. \frac{\partial \boldsymbol{v}}{\partial t} \right|_{\text{in } O'x'y'z'} dQ'$$

$$- \rho \underbrace{\dot{\boldsymbol{r}}_O(t) \times \int_{Q'(t)} \boldsymbol{v} dQ'}_{\boldsymbol{v}_O \times \boldsymbol{M}} + \rho \int_{S'(t)+\Sigma'(t)} (\boldsymbol{r}' - \boldsymbol{r}_O(t)) \times \boldsymbol{v}\, U_n dS'. \quad (2.74)$$

Remembering $\boldsymbol{r} = \boldsymbol{r}' - \boldsymbol{r}_O(t)$ and (2.65) makes it possible to get an alternative expression for the second integral term. Further, we rewrite the first integral on the right-hand side of (2.74) by using the Euler equations and adopting the convective acceleration term. Changing to integration by $Q(t)$ gives

$$\rho \int_{Q(t)} \boldsymbol{r} \times \left.\frac{\partial \boldsymbol{v}}{\partial t}\right|_{\text{in } O'x'y'z'} dQ = -\rho \int_{Q(t)} \boldsymbol{r} \times \nabla \cdot (\boldsymbol{v}\boldsymbol{v})\, dQ$$

$$- \underbrace{\int_{Q(t)} \boldsymbol{r} \times \nabla(p - p_0) dQ}_{\boldsymbol{M}_O(t)} + \underbrace{\rho \int_{Q(t)} \boldsymbol{r} \times \boldsymbol{g}\, dQ}_{M_l \boldsymbol{r}_{lC} \times \boldsymbol{g}},$$

where the second integral is, by definition, the *hydrodynamic moment* (2.63) according to the Gauss theorem. The first integral can be shown to be equal to $-\rho \int_{S(t)+\Sigma(t)} \boldsymbol{r} \times \boldsymbol{v}(\boldsymbol{v} \cdot \boldsymbol{n})\, dS$.

Because $\boldsymbol{v} \cdot \boldsymbol{n} = U_n$ on $S(t) + \Sigma(t)$ and $\boldsymbol{r} = \boldsymbol{r}' - \boldsymbol{r}_O(t)$, it follows that the first integral is equal to minus the last integral on the right-hand side of (2.74). As a consequence, (2.74) gives

$$\boldsymbol{M}_O = M_l\, \boldsymbol{r}_{lC} \times \boldsymbol{g} - \overset{*}{\boldsymbol{G}}_O - \boldsymbol{\omega} \times \boldsymbol{G}_O - \boldsymbol{v}_O \times \boldsymbol{M}. \quad (2.75)$$

Formula (2.75) is true for an inviscid incompressible liquid including the case of rotational flows.

According to the definition of the inertia tensor (2.52), the velocity field is the sum of $\boldsymbol{v}_O + \nabla(\boldsymbol{\omega} \cdot \boldsymbol{\Omega})$ and $\boldsymbol{v}_1$, i.e.,

$$\boldsymbol{v} = \boldsymbol{v}_0 + \nabla(\boldsymbol{\omega} \cdot \boldsymbol{\Omega}) + \boldsymbol{v}_1, \quad (2.76)$$

where $\boldsymbol{v}_O \cdot \boldsymbol{r} + \boldsymbol{\omega} \cdot \boldsymbol{\Omega}$ satisfies the Laplace equation and describes the potential flow of an incompressible liquid in a completely filled tank. Substituting (2.76) into (2.73) gives

$$\boldsymbol{G}_O = M_l\, \boldsymbol{r}_{lC} \times \boldsymbol{v}_O + \boldsymbol{\omega} \cdot \boldsymbol{J}^1 + \rho \int_{Q(t)} \boldsymbol{r} \times \boldsymbol{v}_1 dQ. \quad (2.77)$$

By considering potential flows, ($\boldsymbol{v} = \nabla \Phi$) allows for simplifications in (2.77). Inserting (2.46) into (2.74), using the definition of the Stokes-Joukowski potentials, the Gauss theorem and the Green first identity lead to

Differential, variational, and modal statements

$$\boldsymbol{G}_O(t) = \rho \int_{Q(t)} \boldsymbol{r} \times \boldsymbol{v}_O \, \mathrm{d}Q + \rho \int_{Q(t)} \boldsymbol{r} \times \nabla(\boldsymbol{\omega} \cdot \boldsymbol{\Omega}) \, \mathrm{d}Q + \rho \int_{Q(t)} \boldsymbol{r} \times \nabla\varphi \, \mathrm{d}Q$$

$$= M_l \, \boldsymbol{r}_{lC} \times \boldsymbol{v}_O + \rho \int_{S(t)+\Sigma(t)} (\boldsymbol{\omega} \cdot \boldsymbol{\Omega})(\boldsymbol{r} \times \boldsymbol{n}) \, \mathrm{d}S + \rho \int_{S(t)+\Sigma(t)} (\boldsymbol{r} \times \boldsymbol{n}) \, \varphi \, \mathrm{d}S$$

$$= M_l \, \boldsymbol{r}_{lC} \times \boldsymbol{v}_O + \rho \int_{S(t)+\Sigma(t)} (\boldsymbol{\omega} \cdot \boldsymbol{\Omega}) \frac{\partial \boldsymbol{\Omega}}{\partial n} \, \mathrm{d}S + \rho \int_{S(t)+\Sigma(t)} \frac{\partial \boldsymbol{\Omega}}{\partial n} \varphi \, \mathrm{d}S. \quad (2.78)$$

Modifying the last integral as

$$\rho \int_{S(t)+\Sigma(t)} \frac{\partial \boldsymbol{\Omega}}{\partial n} \varphi \, \mathrm{d}S = \rho \int_{S(t)+\Sigma(t)} \boldsymbol{\Omega} \frac{\partial \varphi}{\partial n} \, \mathrm{d}S = -\rho \int_{\Sigma(t)} \boldsymbol{\Omega} \frac{\partial \xi/\partial t}{|\nabla \xi|} \, \mathrm{d}S$$

$$= \frac{d^*}{dt} \rho \int_{Q(t)} \boldsymbol{\Omega} \, \mathrm{d}Q - \rho \int_{Q(t)} \frac{\partial \boldsymbol{\Omega}}{\partial t} \, \mathrm{d}Q$$

derives

$$\boldsymbol{G}_O = M_l \, \boldsymbol{r}_{lC} \times \boldsymbol{v}_O + \boldsymbol{\omega} \cdot \boldsymbol{J}_0^1 + \overset{*}{\boldsymbol{l}}_{\omega t} - \boldsymbol{l}_{\omega t}, \quad (2.79)$$

where vectors $\boldsymbol{l}_\omega = (l_{1\omega}, l_{2\omega}, l_{3\omega})$ and $\boldsymbol{l}_{\omega t} = (l_{1\omega t}, l_{2\omega t}, l_{3\omega t})$ are defined by (2.55) and depend on $\boldsymbol{\Omega}(x, y, z, t)$.

Accounting for (2.79) and (2.70) gives the Lukovsky formula for the hydrodynamic moment

$$\boldsymbol{M}_O = M_l \boldsymbol{r}_{lC} \times \left(\boldsymbol{g} - \boldsymbol{\omega} \times \boldsymbol{v}_O - \overset{*}{\boldsymbol{v}}_O\right) - \boldsymbol{J}^1 \cdot \overset{*}{\boldsymbol{\omega}} - \overset{*}{\boldsymbol{J}^1} \cdot \boldsymbol{\omega} - \boldsymbol{\omega} \times (\boldsymbol{J}^1 \cdot \boldsymbol{\omega})$$

$$- \overset{**}{\boldsymbol{l}}_\omega + \overset{*}{\boldsymbol{l}}_{\omega t} - \boldsymbol{\omega} \times \left(\overset{*}{\boldsymbol{l}}_\omega - \boldsymbol{l}_{\omega t}\right). \quad (2.80)$$

When the rigid tank is completely filled, (2.80) becomes simpler because $\overset{*}{\boldsymbol{l}}_\omega = \boldsymbol{l}_{\omega t} = 0$ and the inertial tensor is independent of the time

$$\boldsymbol{M}_O^{filled}(t) = M_l \boldsymbol{r}_{lC_0} \times \left(\boldsymbol{g} - \boldsymbol{\omega} \times \boldsymbol{v}_O - \overset{*}{\boldsymbol{v}}_O\right) - \boldsymbol{J}^1 \cdot \overset{*}{\boldsymbol{\omega}} - \boldsymbol{\omega} \times (\boldsymbol{J}^1 \cdot \boldsymbol{\omega}). \quad (2.81)$$

Using (2.75) together with (2.71), one can compute the hydrodynamic moment relative to another body-fixed point P as

$$\boldsymbol{M}_P(t) = \boldsymbol{r}_{PO} \times \boldsymbol{F}(t) + \boldsymbol{M}_O(t), \quad (2.82)$$

where $\boldsymbol{r}_{PO}$ is the radius-vector of O relative to P.

3 Linear modal theory and damping

3.1 LINEAR NATURAL SLOSHING MODES AND FREQUENCIES

3.1.1 THEORY

Small initial perturbations of the contained liquid in an unmovable tank lead to small-magnitude (relative to the horizontal size of the mean free surface) sloshing, which is a linear superposition of standing waves whose patterns and frequencies are the same as the natural sloshing modes and frequencies. Getting the natural sloshing modes (exact or approximate) implies a key task of linear and nonlinear multimodal methods. There exist only a few cases, for which the spectral boundary problem on the natural sloshing modes has exact analytical solutions. Fortunately, the upright circular container belongs to those cases.

Figure 3.1 introduces the tank-fixed Cartesian coordinate system $Oxyz$, as well as the hydrostatic liquid domain Q_0 with the unperturbed free surface Σ_0. Normally (but not necessarily), O coincides with the geometrical centre of the mean free surface Σ_0 and the z-axis is vertical upwards. The free-surface problem, which describes the linear liquid sloshing dynamics, includes the Laplace equation (2.3), the zero-Neumann boundary condition on the wetted tank surface (2.11) (the tank velocities $\boldsymbol{v}_O$ and $\boldsymbol{\omega}$ are zeros), and the linearised free-surface conditions (2.12) and (2.17), which take the form

$$\frac{\partial \Phi}{\partial t} + g\zeta = 0 \text{ and } \frac{\partial \Phi}{\partial z} = \frac{\partial \zeta}{\partial t} \text{ on } \Sigma_0, \tag{3.1}$$

where $z = \zeta(x, y, t)$ describes the small-magnitude free-surface elevations. Using (2.18) deduces the volume conservation condition $\int_{\Sigma_0} \zeta \, dxdy = 0$.

Standing waves are associated with the harmonic function Φ, which satisfies the zero-Neuman condition on the mean wetted tank surface S_0, and is a harmonic by the time t, i.e.,

$$\zeta(x,y,t) = \underbrace{f(x,y)}_{\varphi(x,y,0)} \exp(i\sigma t), \quad \Phi(x,y,z,y) = \frac{ig}{\sigma}\varphi(x,y,z)\exp(i\sigma t), \tag{3.2}$$

where σ is the circular frequency and $i^2 = -1$. Substituting the harmonic function φ by (3.2) into (3.1) leads to the spectral boundary problem

$$\nabla^2 \varphi = 0 \text{ in } Q_0, \quad \frac{\partial \varphi}{\partial n} = 0 \text{ on } S_0, \quad \frac{\partial \varphi}{\partial z} = \kappa\varphi \text{ on } \Sigma_0, \quad \int_{\Sigma_0} \varphi \, dxdy = 0 \tag{3.3}$$

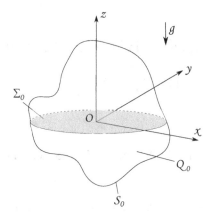

Figure 3.1 The mean liquid domain Q_0, the mean free surface Σ_0, and the mean wetted tank surface S_0 for an unmovable container. When perturbations of the free surface are small relative to the horizontal dimension of Σ_0, the free-surface elevations and internal flows may be described within the framework of the linear sloshing theory.

with the spectral parameter $\kappa = \sigma^2/g$ in the boundary condition on Σ_0.

The mathematical theory of the spectral boundary problem (3.3) is in some detail outlined in [42, 72, 174] and references therein. The corresponding theorems state that the spectrum consists of the positive eigenvalues

$$0 < \kappa_1 \le \kappa_2 \le \ldots \le \kappa_n \le \ldots, \quad \kappa_n \to \infty, \tag{3.4}$$

in which the equalities are due to, possibly, multiple (finite number) eigenfunctions with the same eigenvalue.

The eigenfunctions $\varphi_n(x, y, z)$ are called the natural sloshing modes. The natural sloshing periods and frequencies are expressed in terms of $\{\kappa_n\}$ as

$$T_n = 2\pi/\sqrt{g\kappa_n}, \quad \sigma_n = \sqrt{g\kappa_n}. \tag{3.5}$$

The free-standing wave patterns are determined by

$$z = f_n(x, y) = \varphi_n(x, y, 0) = \frac{1}{\kappa_n}\frac{\partial \varphi_n}{\partial z}(x, y, 0), \tag{3.6}$$

which provide the volume conservation and the orthogonality condition

$$\int_{\Sigma_0} f_i f_j \, dxdy = \int_{\Sigma_0} \varphi_i(x, y, 0)\varphi_j(x, y, 0)\, dxdy = 0 \text{ as } i \ne j. \tag{3.7}$$

Furthermore, functions $\{f_i\}$ constitute a functional basis in $L_2(\Sigma_0)$, i.e., any function $f(x, y)$ satisfying the volume conservation condition can be expanded in the functional series by $\{f_i\}$.

The natural sloshing modes φ_n may behave singularly at the (intersection) contact curve between the mean free surface Σ_0 and the wetted tank surface.

This singular behaviour is especially important when the interior angle along the contact curve is larger than 90°. According to the corresponding theorems in [108,109], the first-order spatial derivatives of φ_n become then singular, and, therefore, provide an infinite velocity field on the contact curve.

3.1.2 EXACT SOLUTIONS AND ASSOCIATED WAVE TYPES

Almost all existing analytical solutions of the spectral boundary problem (3.1) are documented in [62]. Two suitable cases when these solutions exist are upright tanks of rectangular and circular cross sections.

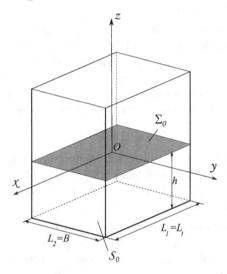

Figure 3.2 Three-dimensional rectangular tank filled with a liquid. Geometric notations for the hydrostatic liquid shape.

3.1.2.1 Rectangular tank

The three-dimensional rectangular tank filled with the mean liquid depth h is depicted in Figure 3.2. Using separation of spatial variables $x, y,$ and z (presenting the analytical solution as $\varphi = X(x)Y(y)Z(z)$) makes it possible to find the following analytical natural sloshing modes

$$\varphi_{ij}(x,\,y,\,z) = \alpha_{ij} f_i^{(1)}(x) f_j^{(2)}(y) \frac{\cosh\left(k_{ij}(z+h)\right)}{\cosh(k_{ij}h)},\ \ i,j \geq 0,\, i^2 + j^2 \neq 0, \quad (3.8)$$

where

$$f_i^{(1)}(x) = \cos\left(i\pi\left(x + \tfrac{1}{2}L_1\right)/L_1\right),\ \ f_j^{(2)}(y) = \cos\left(j\pi\left(y + \tfrac{1}{2}L_2\right)/L_2\right) \quad (3.9)$$

are associated with the two-dimensional standing-wave modes and

$$\sigma_{ij}^2/g = \kappa_{ij} = k_{ij}\tanh(k_{ij}h),\ \ k_{ij} = \pi\sqrt{(i/L_1)^2 + (j/L_2)^2}; \quad (3.10)$$

the non-zero multipliers α_{ij} in (3.8) are usually taken equal to the unit. According to (3.6), the standing waves patterns by (3.8) are defined as

$$z = f_{ij}(x, y) = \alpha_{ij}\varphi_{ij}(x,y,0) = f_i^{(1)}(x)f_j^{(2)}(y). \qquad (3.11)$$

When $ij = 0$, (3.11) determines the two-dimensional *Stokes standing wave profiles* by either $z = af_i^{(1)}(x)$ (in the Ox direction) or $z = bf_j^{(2)}(y)$ (in the Oy direction). These are sometimes called *planar* as if we follow the pendulum analogy and consider the pendulum motions in the Oxz and Oyz planes, respectively. For $ij \neq 0$, (3.11) determines strongly three-dimensional wave patterns including of the saddle-point type for $i = j = 1$.

Special wave types are expected for the aspect ratios L_2/L_1 when the natural sloshing frequencies are multiple. For the square base with $L_2/L_1 = 1$, these are swirling and squares-like (standing) wave modes. Indeed, according to (3.10), $\sigma_{0i} = \sigma_{i0}$, $i \geq 1$, which means that two perpendicular Stokes waves have equal frequencies and, as a consequence, superposition of these perpendicular waves may constitute the *combined free-standing Stokes mode*

$$z = \cos(\sigma_{0i}t + \theta_i)S_i(x,y;a,b) = [af_i^{(1)}(x) + bf_i^{(2)}(y)]\cos(\sigma_{0i}t + \theta_i), \quad (3.12)$$

where θ_i is a given phase lag. Utilising the pendulum analogy and terminology from [55, 91], (3.12) introduces *diagonal* ($|a| = |b|$) and *squares-like* ($|a| \neq |b|$) free-standing waves as $i = 1$. These waves are an attribute of the *resonant sloshing* in the *square base container* [55].

This resonant sloshing also introduces the swirling wave mode, which, using the pendulum analogy, implies a conical-type pendulum motion. In the simplest case, swirling could be defined by the superposition of the Stokes waves with a 90° phase difference in the time, $z = af_1^{(1)}(x)\cos\sigma_{10}t + bf_1^{(2)}(y)\sin\sigma_{01}t$, so that it 'propagates' either counterclockwise ($ab > 0$) or clockwise ($ab < 0$).

In the most general case of the resonant sloshing, the $2\pi/\sigma_{01}$-periodic waves by the two lowest Stokes modes are determined by

$$z = f_1^{(1)}(x)[a\cos\sigma_{01}t + \bar{a}\sin\sigma_{01}t] + f_1^{(2)}(y)[\bar{b}\cos\sigma_{01}t + b\sin\sigma_{01}t]. \quad (3.13)$$

Equation (3.13) implies either standing wave or clockwise/counterclockwise swirling. The wave type is specified by the parameter $\Xi = ab - \bar{a}\bar{b}$, i.e.,

$$\Xi > 0 \text{ (counterclockwise)}, \quad \Xi = 0 \text{ (standing)}, \quad \Xi < 0 \text{ (clockwise)}. \quad (3.14)$$

3.1.2.2 Upright circular tank

After rewriting (3.3) in the cylindrical coordinate system (r, θ, z) ($x = r\cos\theta$, $y = r\sin\theta$, $z = z$), using separation of spatial variables as shown

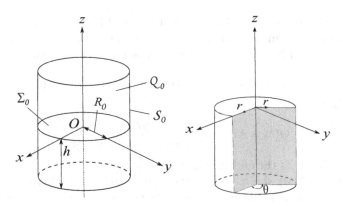

Figure 3.3 Geometric notations for an upright circular cylindrical tank in terms of the circular coordinate system. The hydrostatic liquid shape.

in Figure 3.3 derives the exact analytical solution

$$\varphi_{Mi}(r,\theta,z) = \alpha_{Mi} J_M\left(k_{Mi}\frac{r}{R_0}\right) \frac{\cosh(k_{Mi}(z+h)/R_0)}{\cosh(k_{Mi}h/R_0)} \begin{cases} \cos(M\theta) \\ \sin(M\theta) \end{cases}, \quad (3.15)$$

$M = 0, 1, ...; i = 1, 2, ...$, where α_{Mi} are, as above, any non-zero normalising multipliers, $J_M(\cdot)$ are the Bessel functions of the first kind, and k_{Mi} are the roots (infinite number for each M) of the equation $J'_M(k_{Mi}) = 0$. The surface-wave patterns by the natural sloshing modes (3.15) are defined by

$$z = f_{Mi}(r,\theta) = \alpha_{Mi} \varphi_{Mi}(r,\theta,0) = \alpha_{Mi} J_M\left(k_{Mi}\frac{r}{R_0}\right) \begin{cases} \cos(M\theta) \\ \sin(M\theta) \end{cases}; \quad (3.16)$$

the corresponding natural sloshing frequencies σ_{Mi} come from the expression

$$\sigma_{Mi}^2 R_0/g = R_0 \kappa_{Mi} = k_{Mi} \tanh(k_{Mi}h/R_0). \quad (3.17)$$

The natural sloshing modes (3.15) are parametrised by the two integer indices $M = 0, 1, ...$ and $i = 1, 2, ...$. When $M \neq 0$, the eigenvalues κ_{Mi} (natural frequencies σ_{Mi}) are characterised by the double multiplicity associated with the sine or cosine-multiplier in (3.16). As a consequence, the circular base admits *standing* and *swirling* (rotary) resonant wave modes. These are best exemplified when considering the two lowest anti-symmetric surface modes corresponding to $M = i = 1$ and the lowest natural sloshing frequency σ_{11}. The free-surface profiles can then be defined as

$$z = \alpha_{11} J_1\left(\frac{k_{11}r}{R_0}\right)\left[\cos\theta(a\cos\sigma_{11}t + \bar{a}\sin\sigma_{11}t) + \sin\theta(\bar{b}\cos\sigma_{11}t + b\sin\sigma_{11}t)\right], \quad (3.18)$$

where $a, \bar{a}, \bar{b}, b$ are the wave-amplitude parameters. To explain why (3.18) determines either standing or swirling wave modes, one should consider the so-called nodal line, which, by definition, is formed by intersection of the surface

(3.18) and the mean free surface $z = 0$. The nodal line does not depend on r, and therefore, it is an interval (a straight line) going through the origin. The line is determined by $\theta = \Theta(t)$, $r \leq R_0$, where $\Theta(t)$ comes from

$$\cos\Theta(t)[a\cos\sigma_{11}t + \bar{a}\sin\sigma_{11}t] + \sin\Theta(t)[\bar{b}\cos\sigma_{11}t + b\sin\sigma_{11}t] = 0, \quad (3.19)$$

whose solution is

$$\Theta(t) = \text{Arctan}\left[-\frac{a\cos\sigma_{11}t + \bar{a}\sin\sigma_{11}t}{\bar{b}\cos\sigma_{11}t + b\sin\sigma_{11}t}\right], \quad (3.20)$$

where the Arctan is the multivalued inverse tangent function. Using the time-derivative of (3.20) shows that $\Theta'(t) > 0$ as $ab - \bar{a}\bar{b} > 0$ and $\Theta'(t) < 0$ for $ab - \bar{a}\bar{b} < 0$. This means that $\Theta(t)$ increases and the nodal line rotates counterclockwise for $\Xi = ab - \bar{a}\bar{b} > 0$, but $\Xi < 0$ implies its clockwise rotation. The case $\Xi = 0$ corresponds to the standing-wave mode (the nodal line is unmovable). As matter of the fact, the criterion (3.14) holds true.

3.1.2.3 Annular and sectored upright circular tank

Separating the spatial variables leads to exact analytical natural sloshing modes for annular and sectored circular tank, which is depicted in Figure 3.4. Here $0 < \theta_1 < 2\pi$ is the sector angle, and R_1 and R_2 are the internal and external radii, respectively. The case $R_1 = 0$ corresponds to a non-annular, but, possibly, sectored tank; the zero-Neumann condition at $r = R_1$ should then be excluded. The case $\theta_1 = 2\pi$ corresponds to a non-sectored tank; this also needs omitting boundary conditions at $\theta = 0$ and $\theta = \theta_1$ by changing them to the periodicity condition $\varphi(r, \theta, z) = \varphi(r, \theta + 2\pi, z)$. Section 4.11.3

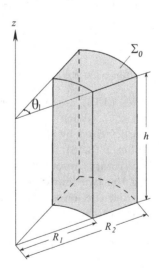

Figure 3.4 Annular and sectored upright circular tank depicted in the cylindrical coordinate system; $R_1 < r < R_2$, $0 < \theta < \theta_1$, $-h < z < 0$.

Linear modal theory and damping

in [62] derives the natural sloshing modes as follows

$$\varphi_{Mi} = \alpha_{Mi} \cos\left(\frac{M\theta}{2\alpha}\right) \frac{\cosh\left(k_{Mi}(z/R_2 + h/R_2)\right)}{\cosh\left(k_{Mi}h/R_2\right)} C_{\frac{M}{2\alpha}}\left(k_{Mi}\frac{r}{R_2}\right),$$

where

$$C_{\frac{M}{2\alpha}}\left(k_{Mi}\frac{r}{R_2}\right) = \det \begin{vmatrix} J_{\frac{M}{2\alpha}}\left(k_{Mi}\frac{r}{R_2}\right) & Y_{\frac{M}{2\alpha}}\left(k_{Mi}\frac{r}{R_2}\right) \\ J'_{\frac{M}{2\alpha}}(k_{Mi}) & Y'_{\frac{M}{2\alpha}}(k_{Mi}) \end{vmatrix}, \; \alpha = \frac{\theta_1}{2\pi},$$

but $J_\mu(\cdot)$ and $Y_\mu(\cdot)$ are the Bessel functions of the first and second kind, respectively; k_{Mi} are the roots of $C'_{\frac{M}{2\alpha}}(k_{Mi}R_1/R_2) = 0$.

3.2 LINEAR SLOSHING PROBLEM AND ITS MODAL SOLUTION

The linear sloshing problem has an analytical solution, which can be found by employing the linear multimodal method. The method gives this analytical solution in terms of the natural sloshing modes φ_n and the Stokes-Joukowski potentials Ω_{0i}. This fact is rarely mentioned in the literature. As a consequence, many researchers continue developing numerical approaches to the linear sloshing problem instead of utilising the analytical solution. This is especially curious when the tank shape admits exact analytical φ_n and Ω_{0i}.

As earlier, we operate with the two coordinate systems, inertial $O'x'y'z'$ and non-inertial (tank-fixed) $Oxyz$. The latter coordinate frame oscillates translatory with a small amplitude and performs angular displacements. The linear sloshing theory suggests that oscillatory tank motions are of a smaller amplitude than the tank size and one can neglect nonlinear terms in the governing equations and boundary conditions. The latter is possible for non-resonant tank excitations.

For brevity, we consider a prescribed tank motion in the inertial coordinate frame $O'x'y'z'$ and use $\dot{\eta}_j(t)$, $j = 1,2,3$ as components of the translational velocity of the origin O, i.e.,

$$\boldsymbol{v}_O(t) = v_{O1}(t)\boldsymbol{e}_1(t) + v_{O2}(t)\boldsymbol{e}_2(t) + v_{O3}(t)\boldsymbol{e}_3(t) = \dot{\eta}_1\boldsymbol{e}_1 + \dot{\eta}_2\boldsymbol{e}_2 + \dot{\eta}_3\boldsymbol{e}_3, \quad (3.21)$$

where $\boldsymbol{e}_i(t)$, $i = 1,2,3$ are the $Oxyz$ coordinate units, which are generally time-dependent. Analogously, the instant angular velocity of the $Oxyz$-system relative to the absolute frame $O'x'y'z'$ reads as

$$\boldsymbol{\omega}(t) = \omega_1(t)\boldsymbol{e}_1(t) + \omega_2(t)\boldsymbol{e}_2(t) + \omega_3(t)\boldsymbol{e}_3(t) = \dot{\eta}_4\boldsymbol{e}_1 + \dot{\eta}_5\boldsymbol{e}_2 + \dot{\eta}_6\boldsymbol{e}_3. \quad (3.22)$$

Remark 3.1. *After introducing (3.21) and (3.22), one can recall Remark 2.3 and assumptions of the linear theory, which neglects all nonlinear terms by $\eta_i(t)$ and their time-derivatives. This implies, in particular, that $\dot{\boldsymbol{v}}_O = \overset{*}{\boldsymbol{v}}_O$ and $\boldsymbol{\omega} = (\dot{\eta}_4, \dot{\eta}_5, \dot{\eta}_6)$, so that $\eta_i(t), i = 1,\ldots,6$ can be interpreted as a small-magnitude perturbation of the $Oxyz$-frame.*

The next focus is the gravity acceleration vector in the body-fixed coordinates, namely, $\boldsymbol{g} = g_1(t)\boldsymbol{e}_1(t) + g_2(t)\boldsymbol{e}_2(t) + g_3(t)\boldsymbol{e}_3(t)$. The linearised projections of the gravity acceleration are expressed as

$$g_1(t) = g\eta_5(t), \quad g_2(t) = -g\eta_4(t), \quad g_3(t) = -g. \tag{3.23}$$

3.2.1 LINEAR MULTIMODAL METHOD

3.2.1.1 Linearised sloshing problem

Let us consider linear perturbations of the free surface relative to its mean position Σ_0 (belongs to Oxy) in the tank-fixed coordinate system, which can then be expressed by the single-valued equation $z = \zeta(x, y, t)$. The boundary-value problem from Chapter 2 should be linearised. The liquid velocity is assumed to have the same order as the tank velocity. Neglecting the nonlinear terms (crossing out the nonlinear quantities in the original free boundary-value problem) after a Taylor expansion of the free-surface condition about Σ_0 causes the linear free-surface conditions formulated on Σ_0 instead of on the instantaneous free-surface position, which is unknown *a priori*. The linearised boundary-value problem for the absolute velocity potential Φ reads then as

$$\frac{\partial^2 \Phi}{\partial x^2} + \frac{\partial^2 \Phi}{\partial y^2} + \frac{\partial^2 \Phi}{\partial z^2} = 0 \text{ in } Q_0, \tag{3.24}$$

$$\frac{\partial \Phi}{\partial n} = (\boldsymbol{v}_O + \boldsymbol{\omega} \times \boldsymbol{r}) \cdot \boldsymbol{n} = \boldsymbol{v}_O \cdot \boldsymbol{n} + \boldsymbol{\omega} \cdot (\boldsymbol{r} \times \boldsymbol{n}) \text{ on } S_0, \tag{3.25}$$

$$\frac{\partial \Phi}{\partial n} = \boldsymbol{v}_O \cdot \boldsymbol{n} + \boldsymbol{\omega} \cdot (\boldsymbol{r} \times \boldsymbol{n}) + \frac{\partial \zeta}{\partial t} \text{ on } \Sigma_0, \tag{3.26}$$

$$\frac{\partial \Phi}{\partial t} - g_1 x - g_2 y - g_3 \zeta = 0 \text{ on } \Sigma_0, \tag{3.27}$$

$$\int_{\Sigma_0} \zeta \, dxdy = 0, \tag{3.28}$$

where Q_0 is the mean liquid domain, S_0 is the mean wetted tank surface, $\boldsymbol{n} = n_1 \boldsymbol{e}_1 + n_2 \boldsymbol{e}_2 + n_3 \boldsymbol{e}_3$ is the outer normal vector, and g_i, $i = 1, 2, 3$ are projections of the gravity acceleration in (3.23). The kinematic free-surface boundary condition (3.26) states that the normal component of the absolute liquid velocity on the mean free surface Σ_0 is equal to the sum of the free-surface velocity $\partial \zeta / \partial t$ relative to the tank and the normal component of the tank velocity on Σ_0. The problem (3.24)–(3.25) requires initial conditions for the free-surface elevation and the velocity potential. Alternatively, the periodicity conditions (steady-state waves) can be adopted.

3.2.1.2 Linear modal equations

The multimodal method introduces the hydrodynamic generalised coordinates $\beta_i(t)$ and velocities $R_i(t)$ by using the functional (Fourier) expansion of the free-surface elevation ζ and the absolute velocity potential Φ

$$\zeta(x,y,t) = \sum_{i=1}^{\infty} \beta_i(t)\varphi_i(x,y,0) = \sum_{i=1}^{\infty} \beta_i(t)f_i(x,y), \qquad (3.29)$$

$$\Phi(x,y,z,t) = \boldsymbol{v}_O(t) \cdot \boldsymbol{r} + \boldsymbol{\omega}(t) \cdot \boldsymbol{\Omega}_0(x,y,z) + \underbrace{\sum_{i=1}^{\infty} R_i(t)\varphi_i(x,y,z)}_{\varphi(x,y,z,t)}, \qquad (3.30)$$

where φ_i, $i \geq 1$ are the natural sloshing modes following from (3.3), but (3.30) includes the Stokes-Joukowski potentials $\Omega_{0i}(x,y,z)$, $i = 1,2,3$, defined by

$$\boldsymbol{\Omega}_0 = \Omega_{01}(x,y,z)\boldsymbol{e}_1 + \Omega_{02}(x,y,z)\boldsymbol{e}_2 + \Omega_{03}(x,y,z)\boldsymbol{e}_3 \qquad (3.31)$$

and coming from the Neumann boundary-value problem

$$\nabla^2 \boldsymbol{\Omega}_0 = 0 \text{ in } Q_0, \quad \frac{\partial \boldsymbol{\Omega}_0}{\partial n} = \boldsymbol{r} \times \boldsymbol{n} \quad \text{on} \quad S_0 + \Sigma_0 \Rightarrow$$

$$\frac{\partial \Omega_{01}}{\partial n} = yn_3 - zn_2, \quad \frac{\partial \Omega_{02}}{\partial n} = zn_1 - xn_3, \quad \frac{\partial \Omega_{03}}{\partial n} = xn_2 - yn_1 \text{ on } S_0 + \Sigma_0. \qquad (3.32)$$

The $\boldsymbol{v}_O(t) \cdot \boldsymbol{r}$ and $\boldsymbol{\omega}(t) \cdot \boldsymbol{\Omega}_0(x,y,z)$ terms in (3.30) take care of the $\boldsymbol{v}_O(t) \cdot \boldsymbol{n}$ and $\boldsymbol{\omega} \cdot (\boldsymbol{r} \times \boldsymbol{n})$ terms in (3.25) and (3.26), respectively, but φ is responsible for the relative liquid velocity.

The natural sloshing modes $\{\varphi_i(x,y,0) = f_i(x,y)\}$ on Σ_0 satisfy the orthogonality conditions (3.7). As long as the linear modal solution (3.29) and (3.30) is introduced, solving (3.24)–(3.28) means finding appropriate $\beta_i(t)$ and $R_i(t)$, $i \geq 1$. The Laplace equation (3.24), the boundary condition (3.25) on the mean wetted tank surface, and the volume conservation condition (3.28) are automatically satisfied due to specific structure of (3.29), (3.30), and special properties of the natural modes. However, we have to satisfy the free-surface conditions (3.26) and (3.27).

Substitution of (3.30) and (3.29) into (3.26) gives

$$\sum_{j=1}^{\infty} R_j \kappa_j \varphi_j(x,y,0) = \sum_{j=1}^{\infty} \dot{\beta}_j f_j(x,y),$$

and, due to the orthogonality condition (3.7),

$$\Rightarrow \dot{\beta}_j = \kappa_j R_j, \ j \geq 1. \qquad (3.33)$$

Accounting for (3.33) in (3.29) and (3.30), as well as definitions (3.21) and (3.22), the dynamic free-surface condition (3.27) leads to

$$\sum_{j=1}^{\infty} \left(\ddot{\beta}_j + \sigma_j^2 \beta_j\right) \kappa_j^{-1} \varphi_j(x,y,0) + x(\ddot{\eta}_1 - g\eta_5) + y(\ddot{\eta}_2 + g\eta_4)$$

$$+ \sum_{k=4}^{6} \ddot{\eta}_k \Omega_{0(k-3)}(x,y,0) = 0 \text{ on } \Sigma_0. \quad (3.34)$$

Multiplying equation (3.34) with $\rho \varphi_m(x,y,0) = \rho f_m(x,y)$, integrating and using the orthogonality condition (3.7) gives the following infinite set of uncoupled linear differential equations for the hydrodynamic generalised coordinates $\{\beta_i\}$ (*linear modal equations*)

$$\mu_m(\ddot{\beta}_m + \sigma_m^2 \beta_m) + \lambda_{1m}(\ddot{\eta}_1 - g\eta_5) + \lambda_{2m}(\ddot{\eta}_2 + g\eta_4)$$

$$+ \sum_{k=4}^{6} \ddot{\eta}_k \lambda_{0(k-3)m} = 0, \ m = 1, 2, \ldots, \quad (3.35)$$

which contain a set of the *hydrodynamic coefficients* to be computed *a priori* by the formulas

$$\sigma_m^2 = g\kappa_m, \ \mu_m = \frac{\rho}{\kappa_m} \int_{\Sigma_0} \varphi_m^2 \, dxdy = \frac{\rho}{\kappa_m} \int_{\Sigma_0} f_m^2 \, dxdy,$$

$$\lambda_{1m} = \rho \int_{\Sigma_0} f_m x \, dxdy, \ \lambda_{2m} = \rho \int_{\Sigma_0} f_m y \, dxdy, \quad (3.36)$$

$$\lambda_{0km} = \rho \int_{\Sigma_0} f_m \Omega_{0k} \, dxdy, \ k = 1, 2, 3; \ m = 1, 2, \ldots.$$

The *index m may be* not only an integer number but also *a pair of integer numbers*, i.e., $m = (ij)$, as it occurs for *three-dimensional tanks* in section 3.1.2. Because the hydrodynamic generalised coordinates $\{\beta_i\}$ do not depend on ρ, it appears unnecessary that ρ is used in the hydrodynamic coefficients. The reason for introducing ρ is that the same hydrodynamic coefficients appear in expressions for the hydrodynamic force and moment.

When the rigid-body motions η_i are known, we rearrange (3.35) to have known variables in the right-hand side, i.e.,

$$\ddot{\beta}_m + \sigma_m^2 \beta_m = K_m(t), \ m = 1, 2, \ldots, \quad (3.37)$$

where the right-hand side reads as

$$K_m(t) = -\frac{\lambda_{1m}}{\mu_m}(\ddot{\eta}_1(t) - g\eta_5(t)) - \frac{\lambda_{2m}}{\mu_m}(\ddot{\eta}_2(t) + g\eta_4(t))$$

$$- \sum_{k=4}^{6} \frac{\ddot{\eta}_k(t) \lambda_{0(k-3)m}}{\mu_m}, \ m = 1, 2, \ldots. \quad (3.38)$$

3.2.1.3 Steady-state and transient waves

The linear sloshing problem describes steady-state and/or transient waves. These follow from (3.29), (3.30), (3.33), and solving the linear modal equations (3.37) and (3.38) instead of solving the boundary-value problem (3.24)–(3.28).

Linear modal theory and damping 43

Steady-state solutions of the linear modal equations (3.37) mean that the forcing terms $K_m(t)$ are T-periodic ($K_m(t+T) = K_m(t)$) and periodic solutions of (3.37) are needed, i.e.,

$$\beta_m(t+T) = \beta_m(t), \quad m = 1, 2, \ldots. \quad (3.39)$$

Transient-wave analysis requires the initial free-surface position $\zeta(x, y, 0)$ and the initial velocity $\dot\zeta(x, y, 0)$ of the free surface. The corresponding initial conditions (2.20) transform then to the Cauchy condition for (3.37)

$$\beta_m(0) = \beta_{0m}, \quad \dot\beta_m(0) = \beta_{1m}, \quad m = 1, 2, \ldots, \quad (3.40)$$

where

$$\beta_{0m} = \frac{\int_{\Sigma_0} f_m \zeta(x,y,0)\,dxdy}{\int_{\Sigma_0} f_m^2\,dxdy}, \quad \beta_{1m} = \frac{\int_{\Sigma_0} f_m \dot\zeta(x,y,0)\,dxdy}{\int_{\Sigma_0} f_m^2\,dxdy}. \quad (3.41)$$

Equations (3.37) have the following general solution:

$$\beta_m(t) = A_m \cos\sigma_m t + B_m \sin\sigma_m t + \sigma_m^{-1}\int_0^t K_m(\tau)\sin(\sigma_m(t-\tau))\,d\tau, \quad (3.42)$$

where constants A_m and B_m may follow from imposing the initial conditions (3.40). When the forcing is harmonic, $K_m(t) = \sigma^2 P'_m \cos\sigma t$ with a non-resonant frequency $\sigma \neq \sigma_m$, solution (3.42) never (physically) becomes periodic. Such a periodic steady-state solution is associated with

$$\beta_m(t) = \frac{P'_m \sigma^2}{\sigma_m^2 - \sigma^2}\cos\sigma t, \quad (3.43)$$

while (3.42) takes then the form

$$\beta_m(t) = a'_m \cos\sigma_m t + b'_m \sin\sigma_m t + \frac{P'_m \sigma^2}{\sigma_m^2 - \sigma^2}\cos\sigma t, \quad (3.44)$$

which becomes $T = 2\pi/\sigma$-periodic, if and only if, $a'_m = b'_m = 0$, namely, when $\beta_m(0) = P'_m\sigma^2/(\sigma_m^2 - \sigma^2)$ and $\dot\beta_m(0) = 0$. This means that the transient solution contains two harmonics associated with the forcing frequency σ and the natural sloshing frequency σ_m. A physical reason for that is that our system has no damping.

There is, in reality, a non-zero *damping* due to the viscous effects. A way to account for the damping is to introduce additional $2\xi_m \sigma_m \dot\beta_m$-terms so that (3.37) takes then the form

$$\ddot\beta_m + 2\xi_m\sigma_m\dot\beta_m + \sigma_m^2\beta_m = K_m(t), \quad m = 1, 2, \ldots. \quad (3.45)$$

Results on harmonically-forced oscillators can be generalized to the case when the modal equations include damping, i.e., for the modal equations (3.45). The general solution takes the form

$$\beta_m(t) = \exp(-\xi_m\sigma_m t)[a_m \cos\sigma'_m t + b_m \sin\sigma'_m t] + c_m \cos\sigma t + d_m \sin\sigma t, \quad (3.46)$$

where

$$\sigma'_m = \sqrt{1-\alpha_m^2}\sigma_m, \quad c_m = \frac{P'_m \sigma^2 \left(\sigma_m^2 - \sigma^2\right)}{[\sigma_m^2 - \sigma^2]^2 + 4\xi_m^2 \sigma_m^2 \sigma^2},$$

$$d_m = \frac{2\alpha_m \sigma_m \sigma^3 P'_m}{[\sigma_m^2 - \sigma^2]^2 + 4\xi_m^2 \sigma_m^2 \sigma^2}.$$

The two coefficients a_m and b_m should be found from the Cauchy conditions (3.40). Obviously, (3.46) tends to the $T = 2\pi/\sigma$-periodic (steady-state) solution on the long-time scale ($t \to \infty$).

3.2.1.4 Lukovsky's formulas for the hydrodynamic force and moment

The book [62] (Chapter 5) presents a detailed derivation of Lukovsky's formulas for the linearised hydrodynamic force and moment. It deals with linearised versions of (2.71) and (2.80), in which (3.29) and (3.30) are inserted. The derivation looks rather interesting and deserving of attention from readers working on sloshing. However, we will not repeat it here, but simply write down the results as they follow from (2.71) and (2.80).

The linearised Lukovsky formula for the hydrodynamic force (2.71) takes the form

$$\boldsymbol{F}(t) = -M_l g \boldsymbol{e}_3 + M_l g(\eta_5 \boldsymbol{e}_1 - \eta_4 \boldsymbol{e}_2) - M_l \left(\overset{*}{\boldsymbol{v}}_O + \dot{\boldsymbol{\omega}} \times \boldsymbol{r}_{lC_0} + \overset{**}{\boldsymbol{r}}_{lC}\right), \quad (3.47)$$

where $\boldsymbol{r}_{lC_0}$ is the hydrostatic position of the liquid mass centre in the $Oxyz$-system, i.e.,

$$\boldsymbol{r}_{lC_0} = \frac{\rho}{M_l} \int_{Q_0} \boldsymbol{r} dQ = x_{lC_0} \boldsymbol{e}_1 + y_{lC_0} \boldsymbol{e}_2 + z_{lC_0} \boldsymbol{e}_3,$$

$$x_{lC_0} = \frac{\rho}{M_l} \int_{Q_0} x dQ, \; y_{lC_0} = \frac{\rho}{M_l} \int_{Q_0} y dQ, \; z_{lC_0} = \frac{\rho}{M_l} \int_{Q_0} z dQ. \quad (3.48)$$

Substituting (3.29) into (3.47) gives the resulting linear hydrodynamic force in terms of the hydrodynamic generalised coordinates

$$\boldsymbol{F}(t) = \underbrace{-M_l g \boldsymbol{e}_3}_{\text{weight}} + \boldsymbol{e}_1 \underbrace{\left[M_l(g\eta_5 - \ddot{\eta}_1 - \ddot{\eta}_5 z_{lC_0} + \ddot{\eta}_6 y_{lC_0}) - \sum_{j=1}^{\infty} \lambda_{1j} \ddot{\beta}_j\right]}_{F_1(t)}$$

$$+ \boldsymbol{e}_2 \underbrace{\left[M_l(-g\eta_4 - \ddot{\eta}_2 - \ddot{\eta}_6 x_{lC_0} + \ddot{\eta}_4 z_{lC_0}) - \sum_{j=1}^{\infty} \lambda_{2j} \ddot{\beta}_j\right]}_{F_2(t)}$$

$$+ \boldsymbol{e}_3 \underbrace{\left[M_l(-\ddot{\eta}_3 - \ddot{\eta}_4 y_{lC_0} + \ddot{\eta}_5 x_{lC_0})\right]}_{F_3(t)}. \quad (3.49)$$

Linear modal theory and damping

When the tank is *completely filled*, the liquid mass centre does not move in the $Oxyz$-frame so that the β_j-dependent quantities vanish and, therefore, (2.72) takes the form

$$\boldsymbol{F}^{filled}(t) = \underbrace{-M_l g \boldsymbol{e}_3}_{\text{weight}} + \boldsymbol{e}_1 \underbrace{[M_l(g\eta_5 - \ddot\eta_1 - \ddot\eta_5 z_{lC_0} + \ddot\eta_6 y_{lC_0})]}_{F_1^{filled}(t)}$$

$$+ \boldsymbol{e}_2 \underbrace{[M_l(-g\eta_4 - \ddot\eta_2 - \ddot\eta_6 x_{lC_0} + \ddot\eta_4 z_{lC_0})]}_{F_2^{filled}(t)} + \boldsymbol{e}_3 \underbrace{[M_l(-\ddot\eta_3 - \ddot\eta_4 y_{lC_0} + \ddot\eta_5 x_{lC_0})]}_{F_3^{filled}(t)},$$

(3.50)

which shows that the linear force acting on the tank surface of the completely filled tank is *the same as the inertial force due to the frozen liquid*.

Neglecting the nonlinear quantities in (2.80) derives the linearised Lukovsky formula for the hydrodynamic moment

$$\boldsymbol{M}_O(t) = -M_l g \boldsymbol{r}_{lC_0} \times \boldsymbol{e}_3 + \Big[M_l g \boldsymbol{r}_{lC_0} \times (\eta_5 \boldsymbol{e}_1 - \eta_4 \boldsymbol{e}_2) - M_l \boldsymbol{r}_{lC_0} \times \overset{*}{\boldsymbol{v}}_O - \boldsymbol{J}_0^1 \cdot \dot{\boldsymbol{\omega}}$$

$$- \rho g \boldsymbol{e}_1 \int_{\Sigma_0} y\zeta(x,y,t) \mathrm{d}S + \rho g \boldsymbol{e}_2 \int_{\Sigma_0} x\zeta(x,y,t) \mathrm{d}S - \rho \int_{\Sigma_0} \boldsymbol{\Omega}_0 \frac{\partial^2 \zeta}{\partial t^2} \mathrm{d}S \Big], \quad (3.51)$$

where the Joukowski inertia tensor $\boldsymbol{J}_0^1$ is introduced, whose elements are

$$J_{0ij}^1 = \rho \int_{S_0 + \Sigma_0} \Omega_{0i} \frac{\partial \Omega_{0j}}{\partial n} \mathrm{d}S. \quad (3.52)$$

Formula (3.51) can be rewritten in terms of the hydrodynamic generalised coordinates:

$$\boldsymbol{M}_O(t) = \underbrace{[M_l g (x_{lC_0} \boldsymbol{e}_2 - y_{lC_0} \boldsymbol{e}_1)]}_{\text{hydrostatic moment}}$$

$$+ \boldsymbol{e}_1 \underbrace{\Big[M_l(g z_{lC_0} \eta_4 + z_{lC_0} \ddot\eta_2 - y_{lC_0} \ddot\eta_3) - \sum_{k=4}^{6} J_{01(k-3)}^1 \ddot\eta_k - \sum_{j=1}^{\infty} \left(g\lambda_{2j}\beta_j + \lambda_{01j}\ddot\beta_j \right) \Big]}_{M_{O1}(t) = F_4(t)}$$

$$+ \boldsymbol{e}_2 \underbrace{\Big[M_l(g z_{lC_0} \eta_5 + x_{lC_0} \ddot\eta_3 - z_{lC_0} \ddot\eta_1) - \sum_{k=4}^{6} J_{02(k-3)}^1 \ddot\eta_k - \sum_{j=1}^{\infty} \left(-g\lambda_{1j}\beta_j + \lambda_{02j}\ddot\beta_j \right) \Big]}_{M_{O2}(t) = F_5(t)}$$

$$+ \boldsymbol{e}_3 \underbrace{\Big[M_l \left(-g(x_{lC_0}\eta_4 + y_{lC_0}\eta_5) + y_{lC_0}\ddot\eta_1 - x_{lC_0}\ddot\eta_2 \right) - \sum_{k=4}^{6} J_{03(k-3)}^1 \ddot\eta_k - \sum_{j=1}^{\infty} \lambda_{03j}\ddot\beta_j \Big]}_{M_{O3}(t) = F_6(t)},$$

(3.53)

and, for the *completely filled tank* ($\beta_j = 0$), equation (2.80) takes the form

$$M_O^{filled}(t) = \underbrace{[M_l g(x_{lC_0}\mathbf{e}_2 - y_{lC_0}\mathbf{e}_1)]}_{\text{hydrostatic moment}}$$

$$+ \mathbf{e}_1 \underbrace{\left[M_l(gz_{lC_0}\eta_4 + z_{lC_0}\ddot{\eta}_2 - y_{lC_0}\ddot{\eta}_3) - \sum_{k=4}^{6} J_{01(k-3)}^1 \ddot{\eta}_k \right]}_{M_{O1}^{filled}(t) = F_4^{filled}(t)}$$

$$+ \mathbf{e}_2 \underbrace{\left[M_l(gz_{lC_0}\eta_5 + x_{lC_0}\ddot{\eta}_3 - z_{lC_0}\ddot{\eta}_1) - \sum_{k=4}^{6} J_{02(k-3)}^1 \ddot{\eta}_k \right]}_{M_{O2}^{filled}(t) = F_5^{filled}(t)}$$

$$+ \mathbf{e}_3 \underbrace{\left[M_l(-g(x_{lC_0}\eta_4 + y_{lC_0}\eta_5) + y_{lC_0}\ddot{\eta}_1 - x_{lC_0}\ddot{\eta}_2) - \sum_{k=4}^{6} J_{03(k-3)}^1 \ddot{\eta}_k \right]}_{M_{O6}^{filled}(t) = F_6^{filled}(t)}.$$

(3.54)

In contrast to the hydrodynamic force, (3.54) shows that the hydrodynamic moment in a completely filled tank *cannot be related to a frozen liquid*.

3.2.2 LINEAR MULTIMODAL THEORY FOR AN UPRIGHT CIRCULAR TANK

3.2.2.1 Hydrodynamic coefficients

Based on the analytically-given natural sloshing modes and Stokes-Joukowski potentials from section 3.1.2.2, one can compute all the hydrodynamic coefficients (3.36) for the case of the vertical circular cylindrical Q_0. The procedure *traditionaly* suggests that the normalising multipliers α_{Mi} in the natural sloshing modes (3.15) provide the unit maximum elevation at the wall by (3.16). Because there are two different modes (associated with cosine and sine multipliers for each natural frequency σ_{Mi}, $M \neq 0$), we need an additional (third) index, i.e.,

$$\varphi_{Mi,1}(r, \theta, z) = R_{Mi}(r) \frac{\cosh(k_{Mi}(z+h)/R_0)}{\cosh(k_{Mi}h/R_0)} \cos(M\theta),$$

$$\varphi_{Mi,2}(r, \theta, z) = R_{Mi}(r) \frac{\cosh(k_{Mi}(z+h)/R_0)}{\cosh(k_{Mi}h/R_0)} \sin(M\theta),$$

(3.55)

where

$$R_{Mi}(r) = \frac{J_M(k_{Mi}r/R_0)}{J_M(k_{Mi})}, \quad \kappa_{Mi} = \frac{\sigma_{Mi}^2}{g} = \frac{k_{Mi}}{R_0} \tanh(k_{Mi}h/R_0) \quad (3.56)$$

Linear modal theory and damping 47

for the chosen normalisation. The surface wave patterns take then the form
$$f_{Mi,1}(r,\theta) = \varphi_{Mi,1}(r,\theta,0) = R_{Mi}(r)\cos(M\theta),$$
$$f_{Mi,2}(r,\theta) = \varphi_{Mi,2}(r,\theta,0) = R_{Mi}(r)\sin(M\theta). \quad (3.57)$$

The Stokes-Joukowski potentials read as
$$\Omega_{01} = -F(r,z)\sin\theta, \quad \Omega_{02} = F(r,z)\cos\theta, \quad \Omega_{03} = 0, \quad (3.58)$$
where (see, [62])
$$F(r,z) = zr - 4R_0^2 \sum_{j=1}^{\infty} \frac{R_{1j}(r)}{(k_{1j}^2-1)k_{1j}} \frac{\sinh\left(k_{1j}(z+\tfrac{1}{2}h)/R_0\right)}{\cosh\left(\tfrac{1}{2}k_{1j}h/R_0\right)}. \quad (3.59)$$

Because the Stokes-Joukowski potentials (3.58) are uniquely defined with $\sin\theta$ and $\cos\theta$ multipliers and $x=r\cos\theta$, $y=r\sin\theta$, the expressions (3.36) lead to the only non-zero hydrodynamic coefficients

$$\mu_{1j,1} = \mu_{1j,2} = \frac{\rho\pi}{\kappa_{1j}}\int_0^{R_0} rR_{1j}^2(r)dr = \frac{\rho\pi R_0^2}{\kappa_{1j}}\frac{k_{1j}^2-1}{2k_{1j}^2} = \frac{\rho\pi R_0^3(k_{1j}^2-1)}{2k_{1j}^3\tanh(k_{1j}h/R_0)},$$

$$\lambda_{1(1j,1)} = \lambda_{2(1j,2)} = \rho\pi\int_0^{R_0} r^2 R_{1j}(r)dr = \frac{\rho\pi R_0^3}{k_{1j}^2},$$

$$\lambda_{02(1j,1)} = -\lambda_{01(1j,2)} = \rho\pi\int_0^{R_0} rR_{1j}(r)F(r,0)dr = -\frac{2\pi\rho R_0^4}{k_{1j}^3}\tanh\left(\frac{k_{1j}h}{2R_0}\right),$$
$$(3.60)$$

but the inertia tensor has only two non-zero diagonal elements
$$J_{011}^1 = J_{022}^1 = \rho\pi R_0^2\left[\tfrac{1}{3}h^3 - \tfrac{3}{4}hR_0^2 + 16R_0^3\sum_{j=1}^{\infty}\frac{\tanh(k_{1j}h/(2R_0))}{k_{1j}^3(k_{1j}^2-1)}\right]. \quad (3.61)$$

3.2.2.2 Modal equations

The modal solution (3.29), (3.30) for this tank shape takes the following form

$$\zeta(r,\theta,t) = \sum_{M=0}^{\infty}\sum_{i=1}^{\infty} R_{Mi}(r)(\cos(M\theta)\,\mathrm{p}_{Mi}(t) + \sin(M\theta)\,\mathrm{r}_{Mi}(t)), \quad (3.62a)$$

$$\Phi(r,\theta,z,t) = \dot\eta_1(t)\,r\cos\theta + \dot\eta_2(t)\,r\sin\theta + F(r,z)[-\dot\eta_4(t)\sin\theta + \dot\eta_5(t)\cos\theta]$$
$$+ \sum_{M=0}^{\infty}\sum_{i=1}^{\infty} R_{Mi}(r)\frac{\cosh(k_{Mi}(z+h)/R_0)}{\cosh(k_{Mi}h/R_0)}(\cos(M\theta)\,\mathrm{P}_{Mi}(t) + \sin(M\theta)\,\mathrm{R}_{Mi}(t)),$$
$$(3.62b)$$

and accounting for (3.38) and explicit expressions (3.60), the modal equations with the non-zero right-hand side are

$$\begin{cases} \ddot{p}_{1j} + \sigma_{1j}^2 p_{1j} = -P_j \left[\ddot{\eta}_1(t) - g\eta_5(t) - S_j \ddot{\eta}_5(t) \right]; \\ \ddot{r}_{1j} + \sigma_{1j}^2 r_{1j} = -P_j \left[\ddot{\eta}_2(t) + g\eta_4(t) + S_j \ddot{\eta}_4(t) \right], j = 1, 2, \dots, \end{cases} \quad (3.63)$$

where

$$P_j = \frac{2k_{1j} \tanh(k_{1j} h/R_0)}{k_{1j}^2 - 1}, \quad S_j = \frac{2R_0 \tanh(k_{1j} h/(2R_0))}{k_{1j}}. \quad (3.64)$$

Other linear modal equations are homogeneous, i.e.,

$$\ddot{p}_{Mj} + \sigma_{Mj}^2 p_{Mj} = 0; \quad \ddot{r}_{mj} + \sigma_{mj}^2 r_{mj} = 0, \quad j = 1, 2, \dots, \quad (3.65)$$

where $M = 0, 2, 3, \dots$ and $m = 2, 3 \dots$.

3.2.2.3 Hydrodynamic force and moment

For the circular base container, the hydrodynamic force components in (3.49) are

$$F_1(t) = M_l \left(g\eta_5 - \ddot{\eta}_1 + \frac{h}{2} \ddot{\eta}_5 \right) - \sum_{j=1}^{\infty} \lambda_{1(1j,1)} \ddot{p}_{1j},$$

$$F_2(t) = M_l \left(-g\eta_4 - \ddot{\eta}_2 - \frac{h}{2} \ddot{\eta}_4 \right) - \sum_{j=1}^{\infty} \lambda_{2(1j,2)} \ddot{r}_{1j}, \quad F_3(t) = -M_l \ddot{\eta}_3, \quad (3.66)$$

where $\lambda_{k(1j,k)}$ are defined in (3.60) and the liquid mass is $M_l = \rho V_l = \rho \pi R_0^2 h$.

Analogously, the non-zero hydrodynamic moment components in (3.53) are

$$F_4(t) = -M_l \frac{h}{2} \left(g\eta_4 + \ddot{\eta}_2 \right) - J_{011}^1 \ddot{\eta}_4 - \sum_{j=1}^{\infty} \left(g\lambda_{2(1j,2)} r_j + \lambda_{01(1j,2)} \ddot{r}_j \right),$$

$$F_5(t) = -M_l \frac{h}{2} \left(g\eta_5 - \ddot{\eta}_1 \right) - J_{022}^1 \ddot{\eta}_5 - \sum_{j=1}^{\infty} \left(-g\lambda_{1(1j,1)} p_j + \lambda_{02(1j,1)} \ddot{p}_j \right), \quad (3.67)$$

where expressions (3.60) and (3.61) determine the hydrodynamic coefficients.

3.3 LINEAR VISCOUS DAMPING

3.3.1 DAMPED HARMONIC OSCILLATOR

Linear vibrating systems whose amplitude decreases over time are frequently called *damped harmonic oscillators*. Since nearly all physical systems involve considerations such as friction, air resistance, pollution (non-ideal nature/system), intermolecular forces, etc., where energy in the system is lost to heat or sound, accounting for the damping is important. Examples of the damped harmonic oscillators include any real oscillatory system like

Linear modal theory and damping

the clock pendulum, liquid sloshing in a cup of coffee or guitar string: after shaking the cup or guitar string vibrating, their amplitude slows down and stops over time.

Considering the damped harmonic oscillators normally assumes that frictional force is proportional to the acting force. By definition, the damped harmonic oscillator is a system that, when displaced from its equilibrium position, experiences a force proportional to the displacement with the Hooke constant k so that if the oscillator consists of a mass m, which pulls the mass in the direction of the point $x = 0$, balance of forces (second Newton's law) reads as

$$-kx(t) - c\dot{x}(t) = m\ddot{x}(t) \quad \Rightarrow \quad \ddot{x}(t) + 2\xi\sigma_0 \dot{x} + \sigma_0^2 x(t) = 0, \qquad (3.68)$$

where c is the damping coefficient, $\sigma_0 = \sqrt{k/m}$ is the undamped eigenfrequency of the oscillator and $\xi = c/(2\sqrt{mk})$ is the damping ratio.

The damping ratio ξ determines the system behaviour. The oscillator can be overdamped ($\xi > 1$), which implies the system returns (exponentially decays) to the steady state without oscillating; if critically damped ($\xi = 1$), the system returns to state position as quickly as possible without oscillating (often desired for doors); or underdamped ($\xi < 1$), the system oscillates with the amplitude gradually decreasing to the state position; the angular frequency of the damped oscillator equals $\sigma_0\sqrt{1-\xi^2}$. Damping in the liquid sloshing dynamics is normally associated with the small nondimensional viscous damping ratio ξ so that the damping effect on the natural sloshing frequencies can be neglected.

Theoretically, specification of the damping ratio (coefficient) is related to the so-called Q-factor, which is defined and linked with the damping ratio as follows

$$Q = 2\pi \frac{\text{energy stored}}{\text{energy dissipated per period } T = 2\pi/\sigma_0} = \frac{2\pi\langle E\rangle}{T\langle \dot{E}\rangle} = \frac{1}{2\xi}, \qquad (3.69)$$

where $\langle E \rangle$ is the total averaged energy and $\langle \dot{E} \rangle$ is the average of energy loss (during the period $T = 2\pi/\sigma_0$).

The method of logarithmic decrement is used for experimental estimates of the damping ratio when $\xi \lesssim 0.5$. The logarithmic decrement is defined as the natural log of the ratio of the amplitude of any successive peaks

$$\delta_{\log} = \frac{1}{n} \ln \frac{x(t)}{x(t+nT)}, \qquad (3.70)$$

where $x(t)$ is the overshoot (amplitude – final value) at the time t and $x(t+nT)$ is the overshoot of the peak n period away. The damping ratio is then found from the logarithmic decrement as

$$\xi = \left(1 + \left(\frac{2\pi}{\delta_{\log}}\right)^2\right)^{-1/2}. \qquad (3.71)$$

3.3.2 VISCOUS LAMINAR BOUNDARY-LAYER AND BULK DAMPING

The damping ratios for the free-standing linear waves (3.8) in rectangular tanks were estimated in [62]. The estimates accounted for the laminar boundary-layer effect on the wetted tank walls (Eq. (6.140) in [62]) and the viscous bulk damping (Eq. (6.55) in [62]). These two dissipative factors are typical for linear and nonlinear liquid sloshing. Moreover, they could be asymptotically estimated within the framework of the model of an ideal liquid with irrotational flows. A reason is that the liquid viscosity generates vortices in relatively small local zones about the laminar viscous layer on the wetted tank surface. The damping may also be affected by more complicated hydrodynamic phenomena like the wave breaking and the dynamic contact line effect. These phenomena typically bear a local character in space, that is, the potential flow component should dominate but the rotational flow component, where it exists, is concentrated in small regions of $Q(t)$.

The damped standing waves in a motionless upright circular container can be modelled by the linear modal equations (3.63) and (3.65):

$$\begin{cases} \ddot{p}_{Mi} + 2\xi_{Mi}\sigma_{Mi}\dot{p}_{Mi} + \sigma^2_{Mi}p_{Mi} = 0, \ M = 0, 1, \ldots, \\ \ddot{r}_{mi} + 2\xi_{mi}\sigma_{mi}\dot{r}_{mi} + \sigma^2_{mi}r_{mi} = 0, \ m = 1, 2, \ldots, \ i = 1, 2, \ldots, \end{cases} \quad (3.72)$$

where the damping ratios ξ_{Mi} express dissipative (viscous) phenomena associated, in particular, with the viscous laminar boundary layer on the wetted tank surface [168], viscous effects on the liquid-tank-gas contact curve [103], the free-surface contamination [91], and, for sufficiently strong perturbations, the wave breaking [203].

When ξ_{Mi} are relatively small, we can assume that the standing wave frequencies equal to the natural sloshing frequencies. According to Henderson and Miles [168], who investigated experimental and theoretical values of ξ_{Mi}, the damping ratios in (3.72) due to the laminar viscous layer on the wetted tank surface and the bulk damping can asymptotically be approximated in terms of the nondimensional *boundary-layer thickness* $\delta_b \ll 1$ or, alternatively, the *Galileo number* [9], Ga $\gg 1$ (the ratio between gravity and viscosity forces),

$$\delta_b = \text{Ga}^{-1/4} = \sqrt{\nu/(g^{1/2}R_0^{3/2})} \ll 1, \quad (3.73)$$

where ν is the kinematic viscosity. Specifically, $\xi^{surf}_{Mi} = O(\delta_b)$ is associated with the laminar boundary layer effect; it can be estimated by using the Keulegan method [103], but $\xi^{bulk}_{Mi} = O(\delta_b^2)$ is related to the bulk viscosity. The sum of ξ^{surf}_{Mi} and ξ^{bulk}_{Mi} gives the lower bound of the entire damping ratios, i.e.,

$$\xi_{Mi} \geq \xi^{(0)}_{Mi} = \xi^{surf}_{Mi} + \xi^{bulk}_{Mi}. \quad (3.74)$$

Following the Q-factor definition (3.69) for harmonic oscillators makes it possible to estimate $\xi^{(0)}_{Mi}$ in (3.74) by computing the average of energy loss of a standing wave by the Mi natural sloshing mode $\langle \dot{E} \rangle_{Mi}$ and the corresponding averaged total energy $\langle E\rangle_{Mi}$ (during the natural sloshing period $T_{Mi} = 2\pi/\sigma_{Mi}$). According to (3.69),

$$\frac{\langle \dot{E}\rangle_{Mi}}{\langle E\rangle_{Mi}} = \frac{4\pi\xi_{Mi}^{(0)}}{T_{Mi}} = 2\xi_{Mi}^{(0)}\sigma_{Mi}. \qquad (3.75)$$

Because the averaged total energy $\langle E\rangle_{Mi}$ equals the double averaged kinetic energy $2\langle E_k\rangle_{Mi}$, using the natural sloshing modes (3.15) computes

$$\langle E\rangle = 2\langle E_k\rangle = \rho\left(\frac{1}{2}R_0^5\sigma_{Mi}^2\underbrace{\int_{Q_0}(\nabla\varphi_{Mi})^2 dQ}_{\text{non-dimensional}}\right)$$

$$= \rho\left(\frac{1}{2}R_0^5\sigma_{Mi}^2\kappa_{Mi}\int_{\Sigma_0}\varphi_{Mi}^2 dS\right) = \rho\alpha_{Mi}^2\left(\frac{1}{2}R_0^5\sigma_{Mi}^2\kappa_{Mi}\Lambda_{MM}\int_0^1 rJ_M^2(k_{Mi}r)dr\right),$$

$$\Lambda_{MM} = \begin{cases} 2\pi, & M = 0; \\ \pi, & M \neq 0. \end{cases}$$

The average of energy loss in (3.75) is the sum of energy losses due to the viscous boundary-layer on the wetted tank surface $\langle \dot{E}_l\rangle$ and the bulk damping $\langle \dot{E}_b\rangle$, i.e.,

$$\langle \dot{E}\rangle = \langle \dot{E}_l\rangle + \langle \dot{E}_b\rangle,$$

for which we have the corresponding computational formulas (Eqs. (24,14) and (25,2) in the Landau and Lifshitz textbook [115]). Applying the first formula in Eq. (24,14) (see also [62, (6.51)]), we arrive at the following expression

$$\langle \dot{E}_l\rangle_{Mi} = \frac{1}{2}\mu\sqrt{\frac{\nu\sigma_{Mi}}{2R_0^2}}R_0^5\sigma_{Mi}^2\underbrace{\int_{S_0}\left|\frac{\partial\varphi_{Mi}}{\partial\tau}\right|^2 dS}_{\text{nondimensional}}, \qquad (3.76)$$

where μ is the coefficient of viscosity, $\nu = \mu/\rho$ is the kinematic viscosity, and $v_{0\tau} = \partial_\tau\varphi_{Mi}/\partial\tau$ is the tangential derivative on S_0.

The wetted tank surface S_0 consists of the bottom S_{0b} and the walls S_{0w}. By splitting the last integral in (3.76) into the corresponding two components gives

$$\int_{S_{0w}}\left|\frac{\partial\varphi_{Mi}}{\partial\tau}\right|^2 dS = \alpha_{Mi}^2 J_M^2(k_{Mi})\frac{\Lambda_{MM}}{2}\left[M^2\left(\frac{\tanh(k_{Mi}\bar{h})}{k_{Mi}} + \frac{\bar{h}}{\cosh^2(k_{Mi}\bar{h})}\right)\right.$$
$$\left. + k_{Mi}^2\left(\frac{\tanh(k_{Mi}\bar{h})}{R_{Mi}} - \frac{\bar{h}}{\cosh^2(k_{Mi}\bar{h})}\right)\right], \qquad (3.77a)$$

$$\int_{S_{0b}}\left|\frac{\partial\varphi_{Mi}}{\partial\tau}\right|^2 dS = \frac{\alpha_{Mi}^2\Lambda_{MM}}{\cosh(k_{Mi}\bar{h})}\left[k_{Mi}^2\int_0^1 r(J_M'(k_{Mi}r))^2 dr + M^2\int_0^1 \frac{J_M^2(k_{Mi}r)}{r}dr\right], \qquad (3.77b)$$

where $\bar{h} = h/R_0$.

Accounting for (3.76)–(3.77) in (3.75), we deduce

$$\xi_{Mi}^{surf} = \delta_b \frac{\mu_{Mi}^{(1)} + \frac{1}{2}J_{Mi}^2(k_{Mi})(\mu_{Mi}^{(2)} + \mu_{Mi}^{(3)})}{2\sqrt{2}\,\kappa_{Mi}^{5/4}\mu_{Mi}^{(0)}}, \qquad (3.78)$$

where

$$\mu_{Mi}^{(1)} = k_{Mi}^2 \int_0^1 r(J'_{Mi}(k_{Mi}r))^2 dr + M^2 \int_0^1 \frac{J_{Mi}^2(k_{Mi}r)}{r} dr,$$

$$\mu_{Mi}^{(2)} = M^2 \left(\frac{\tanh(k_{Mi}\bar{h})}{k_{Mi}} + \frac{\bar{h}}{\cosh^2(k_{Mi}\bar{h})}\right), \quad \mu_{Mi}^{(0)} = \int_0^1 rJ_{Mi}^2(k_{Mi}r)dr, \quad (3.79)$$

$$\mu_{Mi}^{(3)} = k_{Mi}^2 \left(\frac{\tanh(k_{Mi}\bar{h})}{k_{Mi}} - \frac{\bar{h}}{\cosh^2(k_{Mi}\bar{h})}\right).$$

The energy dissipation due to the bulk damping can be found by using the Landau-Lifshitz formula [115], which is rewritten in [168] in the form

$$\langle \dot{E}_b \rangle_{Mi} = \frac{\rho}{2}\frac{\nu}{R_0^2}R_0^5\sigma_{Mi}^2 \int_{S_0+\Sigma_0} \frac{\partial(\nabla\varphi_{Mi})^2}{\partial n} dS. \qquad (3.80)$$

The integral in (3.80) can be found in the analytical form

$$\int_{\Sigma_0+S_0} \frac{\partial(\nabla\varphi_{Mi})^2}{\partial n} dS = -4\alpha_{Mi}^2 \Lambda_{MM}\,\kappa_{Mi}\,k_{Mi}^2 \int_0^1 r(J_M(k_{Mi}r))^2 dr$$

$$+ 2\alpha_{Mi}^2 J_M^2(k_{Mi})M^2\Lambda_{MM}\left[\frac{\tanh(k_{Mi}\bar{h})}{2k_{Mi}} + \frac{\bar{h}}{2\cosh^2(k_{Mi}\bar{h})}\right],$$

therefore,

$$\langle \dot{E}_b \rangle_{Mi} = \frac{1}{2}\rho R_0^3 \sigma_{Mi}^2 \nu \Lambda_{MM} M^2$$

$$\times \left(\frac{\tanh(k_{Mi}\bar{h})}{k_{Mi}} + \frac{\bar{h}}{\cosh^2(k_{Mi}\bar{h})}\right)\left(J_M^2(k_{Mi}) - 4\kappa_{Mi}\int_0^1 rJ_M^2(k_{Mi}r)dr\right).$$

Thus, the second component in (3.74) takes the following form

$$\xi_{Mi}^{bulk} = \delta_b^2 \left[\frac{2k_{Mi}^2}{\kappa_{Mi}^{1/2}} - \frac{J_{Mi}^2(k_{Mi})\,\mu_{Mi}^{(2)}}{2\kappa_{Mi}^{3/2}\mu_{Mi}^{(0)}}\right]. \qquad (3.81)$$

3.4 DAMPING VERSUS SURFACE TENSION

Surface tension and damping were already discussed in the beginning of Chapter 2. Because the viscous boundary layer thickness δ_b by (3.73) and $1/\mathrm{Bo} = T_s/\rho g R_0^2$ increase with decreasing R_0, both the viscous damping and surface tension may become important for relatively small radius R_0. However, our hydrodynamic model and the linear modal theory neglect surface tension. Let us control whether this assumption is correct for realistic input

Linear modal theory and damping 53

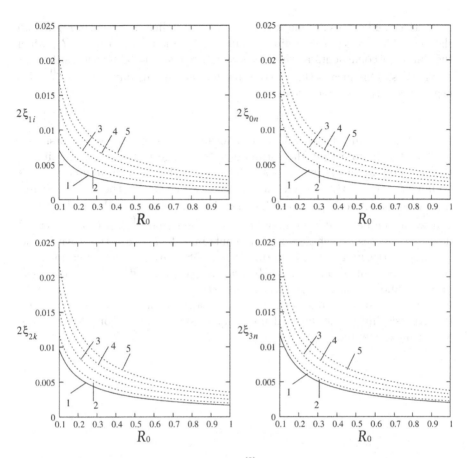

Figure 3.5 Numerical values of $2\xi_{Mi} = 2\xi_{Mi}^{(0)}$ according to (3.74)–(3.79) for $M = 0, 1, 2, 3$. Calculations are done for tap water with the density $\rho = 10^3 \text{ kg/m}^3$, the kinematic viscosity $\nu = 10^{-6} \text{ m}^2/\text{s}$, and $g = 9.81 \text{ m/s}^2$. The nondimensional liquid depth $\bar{h} = 1.5$. The second indices are used to mark the curves. The radius R_0 is measured in meters.

parameters, e.g., for tap water filling the upright circular tank with a finite liquid depth.

Surface tension can be ignored in the liquid sloshing dynamics (away from the contact line) when $100 \lesssim \text{Bo}$ (see Chapter 4 in [62]). For tap water with $\rho = 10^3 \text{ kg/m}^3$, $\nu = 10^{-6} \text{ m}^2/\text{s}$, and $T_s = 0.073 \text{ N/m}$, this estimate computes the lower bound for the tank radius $0.05 \text{ m} \lesssim R_0$ when the developed sloshing theory can be implemented. Is the linear viscous damping by $\xi_{Mi}^{(0)}$ in (3.74) important for $0.05 \text{ m} \lesssim R_0$? The computed damping ratios $\xi_{Mi}^{(0)}$ with (3.78) and (3.81) are shown in Figure 3.5 versus $0.1 \text{ m} < R_0 < 1 \text{ m}$ and the fixed nondimensional liquid depth $\bar{h} = 1.5$. The four panels deal with $M = 0, 1, 2, 3$, respectively. The lines are marked by the second indices $i = 1, ..., 5$. Figure 3.5 shows that the viscous damping is sufficiently small for $0.3\text{ m} \lesssim R_0$, when the damping coefficients are clearly lower of realistic nondimensional forcing amplitudes. This means that accounting for the viscous damping by $\xi_{Mi}^{(0)}$ but neglecting surface tension requires

$$0.05\text{ m} \lesssim R_0 \lesssim 0.3\text{ m} \tag{3.82}$$

for the tank radius. These values are relevant to sloshing in bioreactors.

One should note that the estimate (3.82) suggests that the damping in the hydrodynamic system is basically associated with the laminar viscous boundary-layer at the wetted tank surface, which brings the lowest asymptotic component to $\xi_{Mi}^{(0)}$. These integral damping rates are denoted by ξ_{Mi}, and we remember that they may be much larger that $\xi_{Mi}^{(0)}$ in (3.74). Typical damping sources, which significantly contribute to ξ_{Mi} are the wave breaking and overturning, the free-surface fragmentation and contamination, and the dynamic contact angle effect. Increasing ξ_{Mi} may sufficiently increase the upper bound in (3.82). However, these extra damping sources, except, perhaps, the free-surface contamination, are important for the strongly nonlinear (violent) sloshing. The damped nonlinear sloshing will be considered in the forthcoming chapters.

4 Nonlinear modal theories of nonparametric resonant waves in an upright circular container

4.1 NONLINEAR MODAL EQUATIONS FOR SMALL-AMPLITUDE THREE-DIMENSIONAL TANK MOTIONS

4.1.1 INTRODUCTORY REMARKS

An upright circular base tank of the radius R_0 performs small-amplitude (relative to the radius) motions in the space described by the generalised coordinates $\eta_1(t)$ and $\eta_2(t)$ (responsible for horizontal translatory tank motions), but angular tank motions are represented by $\eta_4(t)$ and $\eta_5(t)$, which are responsible for roll and pitch, respectively. As explained in section 3.2, dealing with the small-amplitude angular forcing means that the nonlinear terms by $\eta_4(t)$ and $\eta_5(t)$ should be neglected. The angular tank motions around Oz (yaw) can not generate sloshing within the framework of the considered inviscid hydrodynamic model assuming no-slip condition on the tank walls. The vertical (parametric, heave) forcing is not considered in this book.

Remark 4.1. *Henceforth, to introduce asymptotic relations between the hydrodynamic generalised coordinates and velocities in the modal systems and the excitation magnitude, the sloshing problem is considered in a dimensionless form. This assumes that the characteristic linear dimension is R_0 (= radius), the characteristic time is $T = 1/\sigma$ (σ is the mean circular forcing frequency), and*

$$g := g/(\sigma^2 R_0), \quad h := h/R_0 \qquad (4.1)$$

are the nondimensional gravity acceleration and the mean liquid depth, respectively.

Remark 4.2. *The nondimensional sloshing problem has the fundamental small parameter $\epsilon \ll 1$, which is associated with the nondimensional excitation, that is,*

$$\eta_i(t) = O(\epsilon), \, i = 1, 2, 4, 5, \qquad (4.2)$$

for the nondimensional sway, surge, pitch, and roll. Furthermore, we will neglect the $o(\epsilon)$-quantities.

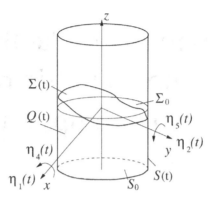

Figure 4.1 The liquid domain $Q(t)$ is bounded by the free surface $\Sigma(t)$ and the wetted tank surface $S(t)$. Liquid sloshing is considered in the noninertial coordinate system $Oxyz$, which is rigidly fixed with the tank so that the coordinate plane Oxy coincides with the mean free surface Σ_0, and Oz is the symmetry axis. The prescribed oscillatory tank motion is described by the generalised coordinates $\eta_1(t)$ (surge), $\eta_4(t)$ (roll), $\eta_2(t)$ (sway), and $\eta_5(t)$ (pitch). The forcing amplitudes of η_1 and η_2 are small relative to the tank radius R_0, but η_4 and η_5 are small nondimensional values on the ϵ-scale. Heave and yaw are not considered. The introduced notations suggest the nondimensional formulation according to Remarks 4.1 and 4.3.

Remark 4.3. *Thoughtful readers may note that previous chapters employed notations $\eta_i(t)$ for the dimensional tank forcing, but the present and forthcoming chapters suggest $\eta_i(t)$ are nondimensional. The same is true for other notations, including for the gravity acceleration and the mean liquid depth in (4.1) as well as for the free surface $\Sigma(t)$, the liquid domain $Q(t)$, and the wetted tank surface $S(t)$, which are demonstrated in Figure 4.1.*

The free surface $\Sigma(t)$ is posed in the single-valued form $z = \zeta(r, \theta, t)$, and the liquid motions inside $Q(t)$ are described by the nondimensional velocity potential $\Phi(r, \theta, z, t)$. The unknowns ζ and Φ are defined in the cylindrical coordinate system (r, θ, z) associated with the already-introduced Cartesian coordinates $Oxyz$. Functions ζ and Φ can be found from either the corresponding boundary value problem or its equivalent variational formulation.

Remark 4.4. *As in the linear sloshing theory, the forthcoming analysis neglects nonlinear terms by $\eta_j(t)$ because they are of the order $o(\epsilon)$ and must be omitted according to Remark 4.2. A particular conclusion is that the nondimensional $\boldsymbol{v}_O$ and $\boldsymbol{\omega}$ are formally defined by (3.21) and (3.22) and keep features of the linear sloshing theory mentioned in Remark 3.1. This includes (3.23) for the nondimensional gravity acceleration $\boldsymbol{g}$ and the 'asymptotic' differentiation rules $\overset{*}{\boldsymbol{v}}_O = \dot{\boldsymbol{v}}_O = \ddot{\eta}_1 \boldsymbol{e}_1 + \ddot{\eta}_2 \boldsymbol{e}_2$ and $\overset{*}{\boldsymbol{\omega}} = \dot{\boldsymbol{\omega}} = \ddot{\eta}_4 \boldsymbol{e}_1 + \ddot{\eta}_5 \boldsymbol{e}_5$ for the absolute nondimensional translational velocity and the* angular instantaneous velocity, respectively. The star- and dot-differentiations are defined in Remark 2.3.

4.1.2 NATURAL SLOSHING MODES AND THEIR NORMALISATION

The Bateman-Luke variational principle is a traditional mathematical tool of the multimodal method, which employs functional decompositions of ζ and Φ where the time-dependent coefficients are interpreted as the hydrodynamic generalised coordinates and velocities, respectively. The decompositions usually adopt the natural sloshing modes as the functional basis. For the upright circular tank shape, the natural sloshing modes and frequencies were derived from (3.3) in section 3.1.2.2. They take the form (3.15)–(3.17).

Following [52,196,197], we write down the natural sloshing modes by using a specific normalisation, which reflects the nondimensional statement and simplifies the variational projective scheme. According to (3.15), the nondimensional natural sloshing modes take the form

$$\varphi_{Mi}(r,z,\theta) = \mathcal{R}_{Mi}(r)\,\mathcal{Z}_{Mi}(z)\genfrac{}{}{0pt}{}{\cos}{\sin}(M\theta), \quad M=0,\ldots;\,i=1,\ldots, \qquad (4.3)$$

where

$$\mathcal{R}_{Mi}(r) = \alpha_{Mi} J_M(k_{Mi}r), \quad \mathcal{Z}_{Mi}(z) = \frac{\cosh(k_{Mi}(z+h))}{\cosh(k_{Mi}h)}, \qquad (4.4)$$

$J_M(\cdot)$ is the Bessel function of the first kind, the radial wave numbers k_{Mi} are determined from the equations $J'_M(k_{Mi}) = 0$ (listed in Table 8.1), and the normalising multipliers α_{Mi} will *not* use the scaling of the linear sloshing theory (see details in section 3.2.2.1) but employ the normalising condition

$$\int_0^1 r\,\mathcal{R}_{Mi}(r)\,\mathcal{R}_{Mj}(r)\,\mathrm{d}r = \delta_{ij}, \quad i,j=1,\ldots, \qquad (4.5)$$

where δ_{ij} is the Kronecker delta. Using this normalisation is a more natural way when operating with the projective scheme below. Because

$$\int_0^1 rJ_M^2(k_{Mi}r)\mathrm{d}r = J_M^2(k_{Mi})\frac{k_{Mi}^2 - m^2}{2k_{Mi}},$$

the normalising multipliers are computed by the formula

$$\alpha_{Mi}^2 = \frac{2k_{Mi}^2}{J_M^2(k_{Mi})(k_{Mi}^2 - M^2)}. \qquad (4.6)$$

The spectral parameter κ_{Mi} and the eigenfrequencies σ_{Mi} are determined by (3.17), which means for the nondimensional statement

$$\kappa_{Mi} = k_{Mi}\tanh(k_{Mi}h), \quad \sigma_{Mi}^2 = \kappa_{Mi}\,g, \quad \bar{\sigma}_{Mi}^2 = \sigma_{Mi}^2/\sigma^2, \qquad (4.7)$$

where g is the nondimensional gravity acceleration and h, as we commented in Remark 4.1, is the R_0-scaled mean liquid depth.

As explained in Chapter 3, the small-amplitude angular tank motions (4.2) require introducing the Stokes-Joukowski potentials $\Omega_{0i}(r,z,\theta)$, $i=1,2,3$, which are harmonic functions satisfying the Neumann conditions (see the dimensional problem (3.32)):

$$\frac{\partial \Omega_{01}}{\partial n} = -(zn_r - rn_z)\sin\theta, \quad \frac{\partial \Omega_{02}}{\partial n} = (zn_r - rn_z)\cos\theta, \quad \frac{\partial \Omega_{03}}{\partial n} = 0 \quad (4.8)$$

on Σ_0 and $S_0 = S_{0b} + S_{0w}$ (see Figure 4.1), where n_r and n_z are the external normal components in the r and z directions ($n_z = 0$ on S_{0w}, but $n_r = 0$ on S_{0b} and Σ_0). The Stokes-Joukowski potentials were derived in section 3.2.2.1 but for dimensional formulation and with other normalising coefficients α_{Mi}. Following these derivations in the considered case, we have

$$\Omega_{01} = -\mathcal{F}(r,z)\sin\theta, \quad \Omega_{02} = \mathcal{F}(r,z)\cos\theta, \quad \Omega_{03} = 0, \quad (4.9)$$

$$\mathcal{F}(r,z) = rz - 2\sum_{n=1}^{\infty} \frac{P_n}{k_{1n}} \mathcal{R}_{1n}(r) \frac{\sinh(k_{1n}(z+\tfrac{1}{2}h))}{\cosh(\tfrac{1}{2}k_{1n}h)},$$

$$P_n = \int_0^1 r^2 \mathcal{R}_{1n}(r)\,dr = \alpha_{1n}\frac{J_1(k_{1n})}{k_{1n}^2}, \quad (4.10)$$

where α_{1n} are computed by (4.6).

4.1.3 MILES-LUKOVSKY MODAL SYSTEM

The nonlinear Miles-Lukovsky modal system (2.59), (2.60) should be rewritten in terms of the generalised coordinates ($p_{Mi}(t)$ and $r_{mi}(t)$) and velocities ($P_{Mi}(t)$ and $R_{mi}(t)$), which are introduced as follows

$$\zeta(r,\theta,t) = \sum_{M,i}^{I_\theta, I_r} \mathcal{R}_{Mi}(r)\cos(M\theta)\,p_{Mi}(t) + \sum_{m,i}^{I_\theta, I_r} \mathcal{R}_{mi}(r)\sin(m\theta)\,r_{mi}(t), \quad (4.11a)$$

$$\Phi(r,\theta,z,t) = \dot{\eta}_1(t)\,r\cos\theta + \dot{\eta}_2(t)\,r\sin\theta + \mathcal{F}(r,z)[-\dot{\eta}_4(t)\sin\theta + \dot{\eta}_5(t)\cos\theta]$$
$$+ \sum_{M,i}^{I_\theta,I_r} \mathcal{R}_{Mi}(r)\,\mathcal{Z}_{Mi}(z)\cos(M\theta)\,P_{Mi}(t) + \sum_{m,i}^{I_\theta,I_r} \mathcal{R}_{mi}(r)\,\mathcal{Z}_{mi}(z)\sin(m\theta)\,R_{mi}(t),$$
$$(4.11b)$$

$I_\theta, I_r \to \infty$. Here and thereafter, the large indices imply summation from zero to I_θ, but small indices mean change from one to I_θ or I_r.

The modal representation (4.11) is similar to (3.62) but $\mathcal{R}$ is *not* identical to R in (3.62) since we deal with the nondimensional statement and other normalising coefficients α_{Mi}. When focusing on the resonant sloshing, the hydrodynamic generalised coordinates and velocities should asymptotically be larger than the tank forcing by (4.2), i.e.,

$$O(\epsilon) \lesssim p_{Mi}(t),\ r_{mi}(t) \quad \text{and} \quad O(\epsilon) \lesssim P_{Mi}(t),\ R_{mi}(t). \quad (4.12)$$

Neglecting the $o(\epsilon)$-order terms in (2.60), the right-hand side in (2.60) becomes linearly dependent on $\eta_i(t)$, $i = 4,5$, as it happens in the linear sloshing theory. At the same time, because of (4.12), we retain the full nonlinearity with respect to the hydrodynamic generalised coordinates and velocities. The result is the Miles-Lukovsky modal system (2.59), (2.60) rewritten as

$$\sum_{M,n}^{I_\theta,I_r} \frac{\partial A^p_{Ab}}{\partial p_{Mn}} \dot{p}_{Mn} + \sum_{m,n}^{I_\theta,I_r} \frac{\partial A^p_{Ab}}{\partial r_{mn}} \dot{r}_{mn} = \sum_{M,n}^{I_\theta,I_r} A^{pp}_{(Ab)(Mn)} P_{Mn} + \sum_{m,n}^{I_\theta,I_r} A^{pr}_{(Ab),(Mn)} R_{mn}, \quad (4.13a)$$

$$\sum_{M,n}^{I_\theta,I_r} \frac{\partial A^r_{ab}}{\partial p_{Mn}} \dot{p}_{Mn} + \sum_{m,n}^{I_\theta,I_r} \frac{\partial A^r_{ab}}{\partial r_{mn}} \dot{r}_{mn} = \sum_{M,n}^{I_\theta,I_r} A^{pr}_{(Mn),(ab)} P_{Mn} + \sum_{m,n}^{I_\theta,I_r} A^{rr}_{(ab)(mn)} R_{mn}, \quad (4.13b)$$

$A = 0, \ldots, I_\theta$; $a = 1, \ldots, I_\theta$; $b = 1, \ldots, I_r$; $I_\theta, I_r \to \infty$ (kinematic subsystem), and

$$\sum_{M,n}^{I_\theta,I_r} \frac{\partial A^p_{Mn}}{\partial p_{Ab}} \dot{P}_{Mn} + \sum_{m,n}^{I_\theta,I_r} \frac{\partial A^r_{mn}}{\partial p_{Ab}} \dot{R}_{mn} + \frac{1}{2} \sum_{ML,nk}^{I_\theta,I_r} \frac{\partial A^{pp}_{(Mn)(Lk)}}{\partial p_{Ab}} P_{Mn} P_{Lk}$$

$$+ \sum_{Ml,nk}^{I_\theta,I_r} \frac{\partial A^{pr}_{(Mn),(lk)}}{\partial p_{Ab}} P_{Mn} R_{lk} + \frac{1}{2} \sum_{ml,nk}^{I_\theta,I_r} \frac{\partial A^{rr}_{(mn)(lk)}}{\partial p_{Ab}} R_{mn} R_{lk} + g\Lambda_{AA} p_{Ab}$$

$$+ (\ddot{\eta}_1 - g\eta_5 - S_b\ddot{\eta}_5)\Lambda_{1A} P_b = 0, \quad (4.14a)$$

$$\sum_{M,n}^{I_\theta,I_r} \frac{\partial A^p_{Mn}}{\partial r_{ab}} \dot{P}_{Mn} + \sum_{m,n}^{I_\theta,I_r} \frac{\partial A^r_{mn}}{\partial r_{ab}} \dot{R}_{mn} + \frac{1}{2} \sum_{ML,nk}^{I_\theta,I_r} \frac{\partial A^{pp}_{(Mn)(Kl)}}{\partial r_{ab}} P_{Mn} P_{Lk}$$

$$+ \sum_{Nl,nk}^{I_\theta,I_r} \frac{\partial A^{pr}_{(Mn),(lk)}}{\partial r_{ab}} P_{Mn} R_{lk} + \frac{1}{2} \sum_{ml,nk}^{I_\theta,I_r} \frac{\partial A^{rr}_{(mn)(lk)}}{\partial r_{ab}} R_{mn} R_{lk} + g\Lambda_{aa} r_{ab}$$

$$+ (\ddot{\eta}_2 + g\eta_4 + S_b\ddot{\eta}_4)\Lambda_{1a} P_b = 0, \quad (4.14b)$$

$A = 0, \ldots, I_\theta$; $a = 1, \ldots, I_\theta$; $b = 1, \ldots, I_{r,}$, $I_\theta, I_r \to \infty$ (dynamic subsystem); the comma between these pairs of indices, $(Ab), (Mn)$, means that pairs do not commutate. The coefficients P_b are defined in (4.10), but

$$S_b = 2\,k_{1b}^{-1} \tanh(k_{1b}h/2), \quad \Lambda_{IJ} = \begin{cases} 2\pi, & I = J = 0, \\ \pi\delta_{IJ}, & \text{otherwise}, \end{cases} \quad (4.15)$$

where δ_{IJ} is the Kronecker delta.

The modal system (4.13), (4.14) contains the following expressions, which are nonlinear functions of the hydrodynamic generalised coordinates

$$A^{pp}_{(Ab)(Mn)} = \int_0^1 \int_{-\pi}^{\pi} r\left[\cos A\theta \cos M\theta\, \mathcal{G}^{(1)}_{(Ab)(Mn)} + \sin A\theta \sin M\theta\, \mathcal{G}^{(2)}_{(Ab)(Mn)}\right] d\theta\, dr,$$

$$A^{rr}_{(ab)(mn)} = \int_0^1 \int_{-\pi}^{\pi} r \left[\sin a\theta \, \sin m\theta \, \mathcal{G}^{(1)}_{(ab)(mn)} + \cos a\theta \, \cos m\theta \, \mathcal{G}^{(2)}_{(ab)(mn)} \right] d\theta \, dr,$$

$$A^{pr}_{(Ab),(mn)} = \int_0^1 \int_{-\pi}^{\pi} r \left[\cos A\theta \, \sin m\theta \, \mathcal{G}^{(1)}_{(Ab)(mn)} - \sin A\theta \, \cos m\theta \, \mathcal{G}^{(2)}_{(Ab)(mn)} \right] d\theta \, dr,$$

$$A^{p}_{Ab} = \int_0^1 \int_{-\pi}^{\pi} r \cos(A\theta) \, \mathcal{G}^{(0)}_{Ab} \, d\theta \, dr, \quad A^{r}_{ab} = \int_{r_1}^1 \int_{-\pi}^{\pi} r \sin(a\theta) \, \mathcal{G}^{(0)}_{Ab} \, d\theta \, dr, \quad (4.16)$$

and

$$\mathcal{G}^{(0)}_{Ab} = \mathcal{R}_{Ab}(r) \int_{-h}^{\zeta} \frac{\cosh(k_{Ab}(z+h))}{\cosh(k_{Ab}h)} \, dz = \mathcal{R}_{Ab}(r) I^{(0)}_{(Ab)},$$

$$\mathcal{G}^{(1)}_{(Ab)(Mn)} = \mathcal{R}'_{Ab}(r) \mathcal{R}'_{Mn}(r) I^{(1)}_{(Ab)(Mn)} + \mathcal{R}_{Ab}(r) \mathcal{R}_{Mn}(r) k_{Ab} k_{Mn} I^{(2)}_{(Ab)(Mn)},$$

$$\mathcal{G}^{(2)}_{(Ab)(Mn)} = AM \, r^{-2} \, \mathcal{R}_{Ab}(r) \mathcal{R}_{Mn}(r) I^{(1)}_{(Ab)(Mn)}; \quad (4.17)$$

$$I^{(1)}_{(Ab)(Mn)} = \int_{-h}^{\zeta} \frac{\cosh(k_{Ab}(z+h)) \cosh(k_{Mn}(z+h))}{\cosh(k_{Ab}h) \cosh(k_{Mn}h)} \, dz,$$

$$I^{(2)}_{(Ab)(Mn)} = \int_{-h}^{\zeta} \frac{\sinh(k_{Ab}(z+h)) \sinh(k_{Mn}(z+h))}{\cosh(k_{Ab}h) \cosh(k_{Mn}h)} \, dz. \quad (4.18)$$

The system has, generally speaking, an infinite dimension, $I_\theta, I_r \to \infty$. If we take finite I_θ and I_r, one can numerically solve the Cauchy problem for (4.13), (4.14), which describes transient waves for a given initial scenario.

Assertion 4.1. *For a small-amplitude sway/surge/pitch/surge (almost) periodic excitation of an upright circular cylindrical tank (see Figure 4.1), introducing the hydrodynamic generalised coordinates and velocities through modal representations (4.11), assuming (4.2) and the resonant sloshing with (4.12), as well as neglecting the $o(\epsilon)$-order quantities in the Miles-Lukovsky equations (2.59), (2.60) lead to the approximate modal equations (4.13), (4.14).*

Henceforth, we mainly concentrate on periodic (steady-state) solutions of approximate modal systems derivable from (4.13), (4.14). The steady-state analysis becomes possible after a series of analytical simplifications using special asymptotic relationships between the hydrodynamic generalised coordinates. In addition, we account for damping in the hydrodynamic system.

4.2 ADAPTIVE MODAL EQUATIONS (A WEAKLY-NONLINEAR MODAL THEORY OF THE THIRD ORDER)

According to (4.2) and (4.12), the hydrodynamic generalised coordinates and velocities are asymptotically small but of a lower asymptotic order than the external tank forcing. When exactly defining this lower asymptotic order and neglecting the $o(\epsilon)$-terms, one can utilise the perturbation theory to simplify

(4.13), (4.14) by keeping only polynomial terms by the hydrodynamic generalised coordinates in the modal equations. Such a reduction from fully-nonlinear to weakly-nonlinear modal system is normally associated with the so-called *adaptive* modal theory.

The third-order adaptive modal theory assumes that the generalised hydrodynamic coordinates and velocities are of the $O(\epsilon^{1/3})$-order, i.e.

$$r_{mi} \sim R_{Mi} \sim P_{Mi} \sim p_{Mi} = O(\epsilon^{1/3}). \tag{4.19}$$

When using (4.19), (4.2) and neglecting the $o(\epsilon)$-order terms, the modal equations (4.13), (4.14) can be converted to a system of ordinary differential equations with respect to the hydrodynamic generalised coordinates $r_{mi}(t)$ and $p_{Mi}(t)$. The procedure of getting these equations is rather complicated. It will be described below.

The first stage implies the Taylor decomposition of $I^{(0)}_{(Ab)}, I^{(1)}_{(Ab)(Mn)}$ and $I^{(2)}_{(Ab)(Mn)}$ by ζ, where $\zeta = O(\epsilon^{1/3})$. Analysis of (2.59), (2.60) shows that $I^{(0)}_{(Ab)}$ must be expanded up to the third order, but $I^{1}_{(Ab)(Mn)}$ and $I^{(2)}_{(Ab)(Mn)}$ require the second-order expansion, that is,

$$I^{(0)}_{(Ab)} = k_{Ab}^{-1}\tanh(k_{Ab}h) + \zeta + \tfrac{1}{2}\kappa_{Ab}\zeta^2 + \tfrac{1}{6}k_{Ab}^2\zeta^3 + \ldots, \tag{4.20a}$$

$$I^{(1)}_{(Ab)(Mn)} = O(1) + \zeta + \tfrac{1}{2}(\kappa_{Ab} + \kappa_{Mn})\zeta^2 + \ldots, \tag{4.20b}$$

$$I^{(2)}_{(Ab)(Mn)} = O(1) + \kappa_{Ab}\kappa_{Mn}\zeta + \tfrac{1}{2}(k_{Ab}^2\kappa_{Mn} + k_{Mn}^2\kappa_{Ab})\zeta^2 + \ldots. \tag{4.20c}$$

Substitution of (4.20b) and (4.20c) into (4.17) gives

$$\mathcal{G}^{(1)}_{(Ab)(Mn)} = O(1) + (\mathcal{R}'_{Ab}\mathcal{R}'_{Mn} + \mathcal{R}_{Ab}\mathcal{R}_{Mn}\kappa_{Ab}\kappa_{Mn})\zeta \\ + \tfrac{1}{2}[(\kappa_{Ab} + \kappa_{Mn})\mathcal{R}'_{Ab}\mathcal{R}'_{Mn} + \mathcal{R}_{Ab}\mathcal{R}_{Mn}(k_{Ab}^2\kappa_{Mn} + k_{Mn}^2\kappa_{Ab})]\zeta^2, \tag{4.21a}$$

$$\mathcal{G}^{(2)}_{(Ab)(Mn)} = O(1) + r^{-2}AM\mathcal{R}_{Ab}\mathcal{R}_{Mn}\zeta + \tfrac{1}{2}r^{-2}AM(\kappa_{Ab} + \kappa_{Mn})\mathcal{R}_{Ab}\mathcal{R}_{Mn}\zeta^2. \tag{4.21b}$$

In the second stage, A^p_{Ab} and A^r_{ab} must be expanded to $O(\epsilon)$ in terms of the generalised coordinates,

$$A^p_{Ab} = \Lambda_{AA,p}A_{b} + \tfrac{1}{2}\sum_{MN,ij}^{I_\theta,I_r}\chi^{pp}_{(Mi)(Nj),(Ab)}p_{Mi}p_{Nj} + \tfrac{1}{2}\sum_{mn,ij}^{I_\theta,I_r}\chi^{rr}_{(mi)(nj),(Ab)}r_{mi}r_{nj}$$

$$+\tfrac{1}{3}\sum_{MNK,ijl}^{I_\theta,I_r}\chi^{ppp}_{(Mi)(Nj)(Kl),(Ab)}p_{Mi}p_{Nj}p_{Kl} + \sum_{Mnk,ijl}^{I_\theta,I_r}\chi^{prr}_{(Mi),(nj)(kl),(Ab)}p_{Mi}r_{nj}r_{kl}, \tag{4.22a}$$

$$A^r_{ab} = \Lambda_{,aa}r_{ab} + \sum_{Mn,ij}^{I_\theta,I_r}\chi^{pr}_{(Mi),(nj),(ab)}p_{Mi}r_{nj} + \tfrac{1}{3}\sum_{mnk,ijl}^{I_\theta,I_r}\chi^{rrr}_{(mi)(nj)(kl),(ab)}r_{mi}r_{nj}r_{kl}$$

$$+ \sum_{MNk,ijl}^{I_\theta, I_r} \chi^{ppr}_{(Mi)(Nj),(kl),(ab)} p_{Mi} p_{Nj} r_{kl}, \quad I_\theta, I_r \to \infty, \quad (4.22b)$$

where

$$\chi^{pp}_{(Mi)(Nj),(Ab)} = \kappa_{Ab} \Lambda_{AMN,} \lambda_{(Ab)(Mi)(Nj)},$$
$$\chi^{rr}_{(Mi)(nj),(Ab)} = \kappa_{Ab} \Lambda_{A,mn} \lambda_{(Ab)(mi)(nj)},$$
$$\chi^{ppp}_{(Mi)(Nj)(Kl),(Ab)} = \tfrac{1}{2} k^2_{Ab} \Lambda_{AMNK,} \lambda_{(Ab)(Mi)(Nj)(Kl)},$$
$$\chi^{prr}_{(Mi),(nj)(kl),(Ab)} = \tfrac{1}{2} k^2_{Ab} \Lambda_{AM,nk} \lambda_{(Ab)(Mi)(nj)(kl)},$$
$$\chi^{pr}_{(Mi),(nj),(ab)} = \kappa_{ab} \Lambda_{M,an} \lambda_{(Mi)(nj)(ab)},$$
$$\chi^{rrr}_{(mi)(nj)(kl),(ab)} = \tfrac{1}{2} k^2_{ab} \Lambda_{,amnk} \lambda_{(mi)(nj)(kl)(ab)},$$
$$\chi^{ppr}_{(Mi)(Nj),(kl),(ab)} = \tfrac{1}{2} k^2_{ab} \Lambda_{MN,ak} \lambda_{(Mi)(Nj)(kl)(ab)},$$

and the tensors

$$\Lambda_{M...N,i...j} = \int_{-\pi}^{\pi} [\cos(M\theta)...\cos(N\theta)] \cdot [\sin(i\theta)...\sin(j\theta)] d\theta. \quad (4.23)$$

can be computed by using (4.15) and the recursive formulas

$$\Lambda_{M,i} = 0; \quad \Lambda_{MN,} = \Lambda_{MN}; \quad \Lambda_{,ij} = \pi \delta_{ij} \mathrm{sgn}(ij),$$
$$\Lambda_{M...NK,i...j} = \tfrac{1}{2}(\Lambda_{M...|N-K|,i...j} + \Lambda_{M...|N+K|,i...j}),$$
$$\Lambda_{M...N,i...ljk} = \tfrac{1}{2}(\Lambda_{M...,|j-k|,i...l} - \Lambda_{M...,|j+k|,i...l}),$$

but the tensors λ are determined by the formula

$$\lambda_{(Ab)...(Mn)} = \int_0^1 r \mathcal{R}_{Ab}(r)...\mathcal{R}_{Mn}(r) dr. \quad (4.24)$$

The partial derivatives of (4.22) by the generalised coordinates acquire the form

$$\frac{\partial A^p_{Ab}}{\partial p_{Df}} = \Lambda_{AD,} \delta_{bf} + \sum_{M,i}^{I_\theta,I_r} \chi^{pp}_{(Mi)(Df),(Ab)} p_{Mi} + \sum_{NK,jl}^{I_\theta,I_r} \chi^{ppp}_{(Df)(Nj)(Kl),(Ab)} p_{Nj} p_{Kl}$$

$$+ \sum_{nk,jl}^{I_\theta,I_r} \chi^{prr}_{(Df),(nj)(kl),(Ab)} r_{nj} r_{kl}, \quad (4.25a)$$

$$\frac{\partial A^p_{Ab}}{\partial r_{df}} = \sum_{m,i}^{I_\theta,I_r} \chi^{rr}_{(mi)(df),(Ab)} r_{mi} + 2 \sum_{Mn,ij}^{I_\theta,I_r} \chi^{prr}_{(Mi),(nj)(df),(Ab)} p_{Mi} r_{nj}, \quad (4.25b)$$

$$\frac{\partial A^r_{ab}}{\partial p_{Df}} = \sum_{n,j}^{I_\theta,I_r} \chi^{pr}_{(Df),(nj),(ab)} r_{nj} + 2 \sum_{Mn,ij}^{I_\theta,I_r} \chi^{ppr}_{(Mi)(Df),(nj),(ab)} p_{Mi} r_{nj}, \quad (4.25c)$$

Nonlinear modal theories of nonparametric resonant waves 63

$$\frac{\partial A^r_{ab}}{\partial r_{df}} = \Lambda_{,ad}\delta_{bf} + \sum_{M,i}^{I_\theta,I_r} \chi^{pr}_{(Mi),(df),(ab)} p_{Mi} + \sum_{mn,ij}^{I_\theta,I_r} \chi^{rrr}_{(mi)(nj)(df),(ab)} r_{mi} r_{nj}$$

$$+ \sum_{MN,ij}^{I_\theta,I_r} \chi^{ppr}_{(Mi)(Nj),(df),(ab)} p_{Mi} p_{Nj}, \quad I_\theta, I_r \to \infty. \quad (4.25d)$$

The third stage leads to analytic expressions for A^{pp}_{AbMn}, $A^{rr}_{(ab)(mn)}$ and $A^{pr}_{(Ab),(mn)}$, where only second-order terms are kept. The terms of $O(1)$-order can be taken from the linear modal theory. The result is

$$A^{pp}_{(Ab)(Mn)} = \Lambda_{AM,}\delta_{bn}\kappa_{Ab} + \sum_{K,l}^{I_\theta,I_r} \Pi^{p,p}_{(Kl),(Ab)(Mn)} p_{Kl}$$

$$+ \sum_{KC,ld}^{I_\theta,I_r} \Pi^{p,pp}_{(Kl)(Cd),(Ab)(Mn)} p_{Kl} p_{Cd} + \sum_{kc,ld}^{I_\theta,I_r} \Pi^{p,rr}_{(kl)(cd),(Ab)(Mn)} r_{kl} r_{cd}, \quad (4.26a)$$

$$A^{rr}_{(ab)(mn)} = \Lambda_{,am}\delta_{bn}\kappa_{ab} + \sum_{K,l}^{I_\theta,I_r} \Pi^{r,p}_{(Kl),(ab)(mn)} p_{Kl}$$

$$+ \sum_{KC,ld}^{I_\theta,I_r} \Pi^{r,pp}_{(Kl)(Cd),(ab)(mn)} p_{Kl} p_{Cd} + \sum_{kc,ld}^{I_\theta,I_r} \Pi^{r,rr}_{(kl)(cd),(ab)(mn)} r_{kl} r_{cd}, \quad (4.26b)$$

$$A^{pr}_{(Ab),(mn)} = \sum_{k,l}^{I_\theta,I_r} \Pi^{r}_{(kl),(Ab),(mn)} r_{kl} + \sum_{Kc,ld}^{I_\theta,I_r} \Pi^{pr}_{(Kl),(cd),(Ab),(mn)} p_{Kl} r_{cd}, \quad (4.26c)$$

where

$$\Pi^{p,p}_{(Kl),(Ab)(Mn)} = \Lambda_{AMK,} G^{(11)}_{(Ab)(Mn),(Kl)} + \Lambda_{K,AM} G^{(12)}_{(Ab)(Mn),(Kl)},$$

$$\Pi^{r,p}_{(Kl),(ab)(mn)} = \Lambda_{K,am} G^{(11)}_{(ab)(mn),(Kl)} + \Lambda_{amK,} G^{(12)}_{(ab)(mn),(Kl)},$$

$$\Pi^{r}_{(kl),(Ab),(mn)} = \Lambda_{A,mk} G^{(11)}_{(Ab)(mn),(kl)} - \Lambda_{m,Ak} G^{(12)}_{(Ab)(mn),(kl)},$$

$$\Pi^{p,pp}_{(Kl)(Cd),(Ab)(Mn)} = \Lambda_{AMKC,} G^{(21)}_{(Ab)(Mn),(Kl)(Cd)} + \Lambda_{KC,AM} G^{(22)}_{(Ab)(Mn),(Kl)(Cd)},$$

$$\Pi^{p,rr}_{(kl)(cd),(Ab)(Mn)} = \Lambda_{AM,kc} G^{(21)}_{(Ab)(Mn),(kl)(cd)} + \Lambda_{,AMkc} G^{(22)}_{(Ab)(Mn),(kl)(cd)},$$

$$\Pi^{r,pp}_{(Kl)(Cd),(ab)(mn)} = \Lambda_{KC,am} G^{(21)}_{(ab)(mn),(Kl)(Cd)} + \Lambda_{KCam,} G^{(22)}_{(ab)(mn),(Kl)(Cd)},$$

$$\Pi^{r,rr}_{(kl)(cd),(ab)(mn)} = \Lambda_{,amkc} G^{(21)}_{(ab)(mn),(kl)(cd)} + \Lambda_{am,kc} G^{(22)}_{(ab)(mn),(kl)(cd)},$$

$$\Pi^{pr}_{(Kl),(cd),(Ab),(mn)} = 2[\Lambda_{AK,mc} G^{(21)}_{(Ab)(mn),(Kl)(cd)} - \Lambda_{Km,Ac} G^{(22)}_{(Ab)(mn),(Kl)(cd)}];$$

$$G^{(11)}_{(Ab)(Mn),(Kl)} = \lambda'_{(Ab)(Mn),(Kl)} + \kappa_{Ab}\kappa_{Mn}\lambda_{(Ab)(Mn)(Kl)},$$

$$G^{(12)}_{(Ab)(Mn),(Kl)} = AM\bar{\lambda}_{(Ab)(Mn)(Kl)},$$

$$G^{(21)}_{(Ab)(Mn),(Kl)(Cd)} = \tfrac{1}{2}[(\kappa_{Ab}+\kappa_{Mn})\lambda'_{(Ab)(Mn),(Kl)(Cd)} + (k^2_{Ab}\kappa_{Mn}$$
$$+ k^2_{Mn}\kappa_{Ab})\lambda_{(Ab)(Mn)(Kl)(Cd)}],$$
$$G^{(22)}_{(Ab)(Mn),(Kl)(Cd)} = \tfrac{1}{2}AM(\kappa_{Ab}+\kappa_{Mn})\bar{\lambda}_{(Ab)(Mn)(Kl)(Cd)}$$

and

$$\lambda'_{(Ab)(Mn),(Cd)\ldots(Ef)} = \int_0^1 r \mathcal{R}'_{Ab}(r)\mathcal{R}'_{Mn}(r)\cdot\mathcal{R}_{Cd}(r)\ldots\mathcal{R}_{Ef}(r)\,\mathrm{d}r,$$
$$\bar{\lambda}_{(Ab)\ldots(Mn)} = \int_0^1 r^{-1}\mathcal{R}_{Ab}(r)\ldots\mathcal{R}_{Mn}(r)\,\mathrm{d}r. \qquad (4.27)$$

The partial derivatives of (4.26) by the generalised coordinates read as

$$\frac{\partial A^{pp}_{(Ab)(Cd)}}{\partial p_{Ef}} = \Pi^{p,p}_{(Ef),(Ab)(Cd)} + 2\sum_{M,i}^{I_\theta,I_r}\Pi^{p,pp}_{(Mi)(Ef),(Ab)(Cd)}p_{Mi}, \qquad (4.28a)$$

$$\frac{\partial A^{pp}_{(Ab)(Cd)}}{\partial r_{ef}} = 2\sum_{m,i}^{I_\theta,I_r}\Pi^{p,rr}_{(mi)(ef),(Ab)(Cd)}r_{mi}, \qquad (4.28b)$$

$$\frac{\partial A^{rr}_{(ab)(cd)}}{\partial p_{Ef}} = \Pi^{r,p}_{(Ef),(ab)(cd)} + 2\sum_{M,i}^{I_\theta,I_r}\Pi^{r,pp}_{(Mi)(Ef),(ab)(cd)}p_{Mi}, \qquad (4.28c)$$

$$\frac{\partial A^{rr}_{(ab)(cd)}}{\partial r_{ef}} = 2\sum_{m,i}^{I_\theta,I_r}\Pi^{r,rr}_{(mi)(ef),(ab)(cd)}r_{mi}, \qquad (4.28d)$$

$$\frac{\partial A^{pr}_{(Ab),(cd)}}{\partial p_{Ef}} = \sum_{n,j}^{I_\theta,I_r}\Pi^{pr}_{(Ef),(nj),(Ab)(cd)}r_{nj}, \qquad (4.28e)$$

$$\frac{\partial A^{pr}_{(Ab),(cd)}}{\partial r_{ef}} = \Pi^{r}_{(ef),(Ab),(cd)} + \sum_{M,i}^{I_\theta,I_r}\Pi^{pr}_{(Mi),(ef),(Ab)(cd)}p_{Mi}. \qquad (4.28f)$$

In the fourth stage, the kinetic equations (4.13) must be resolved with respect to the generalised velocities P_{Ab} and R_{ab}. Postulating

$$P_{Ab} = \frac{1}{\kappa_{Ab}}\dot{p}_{Ab} + \sum_{MN,ij}^{I_\theta,I_r} V^{pp}_{(Mi),(Nj),(Ab)}\dot{p}_{Mi}p_{Nj} + \sum_{mn,ij}^{I_\theta,I_r} V^{rr}_{(mi),(nj),(Ab)}\dot{r}_{mi}r_{nj}$$
$$+ \sum_{Mnk,ijl}^{I_\theta,I_r} V^{prr}_{(Mi),(nj),(kl),(Ab)}\dot{p}_{Mi}r_{nj}r_{kl} + \sum_{MNK,ijl}^{I_\theta,I_r} V^{ppp}_{(Mi),(Nj),(Kl),(Ab)}\dot{p}_{Mi}p_{Nj}p_{Kl}$$
$$+ \sum_{Mnk,ijl}^{I_\theta,I_r} V^{rpr}_{(nj),(Mi),(kl),(Ab)}\dot{r}_{nj}p_{Mi}r_{kl}, \qquad (4.29a)$$

$$R_{ab} = \frac{1}{\kappa_{ab}}\dot{r}_{ab} + \sum_{Mn,ij}^{I_\theta,I_r} V^{pr}_{(Mi),(nj),(ab)}\dot{p}_{Mi}r_{nj} + \sum_{Mn,ij}^{I_\theta,I_r} V^{rp}_{(nj),(Mi),(ab)}\dot{r}_{nj}p_{Mi}$$

$$+ \sum_{mnk,ijl}^{I_\theta,I_r} V^{rrr}_{(mi),(nj),(kl),(ab)}\dot{r}_{mi}r_{nj}r_{kl} + \sum_{MNk,ijl}^{I_\theta,I_r} V^{rpp}_{(kl),(Mi),(Nj),(ab)}\dot{r}_{kl}p_{Mi}p_{Nj}$$

$$+ \sum_{MNk,ijl}^{I_\theta,I_r} V^{ppr}_{(Mi),(Nj),(kl),(ab)}\dot{p}_{Mi}p_{Nj}r_{kl}, \quad (4.29b)$$

substituting (4.29) into the kinetic subsystem (4.13), and gathering similar asymptotic quantities derives the formulas for coefficients in (4.29),

$$V^{pp}_{(Mi),(Nj),(Ab)} = \frac{1}{\Lambda_{AA}\kappa_{Ab}}\left[\chi^{pp}_{(Nj)(Mi),(Ab)} - \frac{\Pi^{p,p}_{(Nj),(Ab)(Mi)}}{\kappa_{Mi}}\right],$$

$$V^{rr}_{(mi),(nj),(Ab)} = \frac{1}{\Lambda_{AA}\kappa_{Ab}}\left[\chi^{rr}_{(nj)(mi),(Ab)} - \frac{\Pi^{r}_{(nj),(Ab),(mi)}}{\kappa_{mi}}\right],$$

$$V^{rp}_{(nj),(Mi),(ab)} = \frac{1}{\Lambda_{aa}\kappa_{ab}}\left[\chi^{pr}_{(Mi),(nj),(ab)} - \frac{\Pi^{r,p}_{(Mi),(ab)(nj)}}{\kappa_{nj}}\right],$$

$$V^{pr}_{(Mi),(nj),(ab)} = \frac{1}{\Lambda_{aa}\kappa_{ab}}\left[\chi^{pr}_{(Mi),(nj),(ab)} - \frac{\Pi^{r}_{(nj),(Mi),(ab)}}{\kappa_{Mi}}\right],$$

$$V^{ppp}_{(Mi),(Nj),(Kl),(Ab)} = \frac{1}{\Lambda_{AA}\kappa_{Ab}}\left[\chi^{ppp}_{(Mi)(Nj)(Kl),(Ab)} - \frac{\Pi^{p,pp}_{(Nj)(Kl),(Ab)(Mi)}}{\kappa_{Mi}}\right.$$

$$\left. - \sum_{C,d}^{I_\theta,I_r} V^{pp}_{(Mi),(Nj),(Cd)}\Pi^{p,p}_{(Kl),(Ab)(Cd)}\right],$$

$$V^{prr}_{(Mi),(nj),(kl),(Ab)} = \frac{1}{\Lambda_{AA}\kappa_{Ab}}\left[\chi^{prr}_{(Mi),(nj)(kl),(Ab)} - \frac{\Pi^{p,rr}_{(nj)(kl),(Ab)(Mi)}}{\kappa_{Mi}}\right.$$

$$\left. - \sum_{c,d}^{I_\theta,I_r} V^{pr}_{(Mi),(nj),(cd)}\Pi^{r}_{(kl),(Ab),(cd)}\right],$$

$$V^{rrr}_{(mi),(nj),(kl),(ab)} = \frac{1}{\Lambda_{aa}\kappa_{ab}}\left[\chi^{rrr}_{(nj)(kl)(mi),(ab)} - \frac{\Pi^{r,rr}_{(nj)(kl),(ab)(mi)}}{\kappa_{mi}}\right.$$

$$\left. - \sum_{C,d}^{I_\theta,I_r} V^{rr}_{(mi),(nj),(Cd)}\Pi^{r}_{(kl),(Cd),(ab)}\right],$$

$$V^{rpp}_{(kl),(Mi),(Nj),(ab)} = \frac{1}{\Lambda_{aa}\kappa_{ab}}\left[\chi^{ppr}_{(Mi)(Nj),(kl),(ab)} - \frac{\Pi^{r,pp}_{(Mi)(Nj),(ab)(kl)}}{\kappa_{kl}}\right.$$
$$\left. - \sum_{c,d}^{I_\theta,I_r} V^{rp}_{(kl),(Mi),(cd)}\Pi^{r,p}_{(Nj),(ab),(cd)}\right],$$

$$V^{ppr}_{(Mi),(Nj),(kl),(ab)} = \frac{1}{\Lambda_{aa}\kappa_{ab}}\left[2\chi^{ppr}_{(Mi)(Nj),(kl),(ab)} - \frac{\Pi^{pr}_{(Nj),(kl),(Mi)(ab)}}{\kappa_{Mi}}\right.$$
$$\left. - \sum_{C,d}^{I_\theta,I_r} V^{pp}_{(Mi),(Nj),(Cd)}\Pi^{r}_{(kl),(Cd),(ab)} - \sum_{c,d}^{I_\theta,I_r} V^{pr}_{(Mi),(kl),(cd)}\Pi^{r,p}_{(Nj),(ab)(cd)}\right],$$

$$V^{rpr}_{(nj),(Mi),(kl),(Ab)} = \frac{1}{\Lambda_{AA}\kappa_{Ab}}\left[2\chi^{prr}_{(Mi),(kl)(nj),(Ab)} - \frac{\Pi^{pr}_{(Mi),(kl),(Ab)(nj)}}{\kappa_{nj}}\right.$$
$$\left. - \sum_{C,d}^{I_\theta,I_r} V^{rr}_{(nj),(kl),(Cd)}\Pi^{p,p}_{(Mi),(Ab)(Cd)} - \sum_{c,d}^{I_\theta,I_r} V^{rp}_{(nj),(Mi),(cd)}\Pi^{r}_{(kl),(Ab),(cd)}\right].$$

In the fifth stage, expressions (4.25), (4.28), and (4.29) are substituted into the dynamic modal equation (4.14). By excluding the $o(\epsilon)$-order terms, we obtain the *adaptive nonlinear modal equations*

$$\sum_{M,i}^{I_\theta,I_r}\ddot{p}_{Mi}\left[\delta_{ME}\delta_{if} + \sum_{N,j}^{I_\theta,I_r}d^{pp,(Ef)}_{(Mi),(Nj)}p_{Nj} + \sum_{NK,jl}^{I_\theta,I_r}d^{ppp,(Ef)}_{(Mi),(Nj),(Kl)}p_{Nj}p_{Kl}+\right.$$
$$\left.+\sum_{nk,jl}^{I_\theta,I_r}d^{prr,(Ef)}_{(Mi),(nj),(kl)}r_{nj}r_{kl}\right] + \sum_{mn,ij}^{I_\theta,I_r}\ddot{r}_{mi}r_{nj}\left[d^{rr,(Ef)}_{(mi),(nj)} + \sum_{K,l}^{I_\theta,I_r}d^{rrp,(Ef)}_{(mi),(nj),(Kl)}p_{Kl}\right]$$
$$+\sum_{MN,ij}^{I_\theta,I_r}\dot{p}_{Mi}\dot{p}_{Nj}\left[t^{pp,(Ef)}_{(Mi),(Nj)} + \sum_{K,l}^{I_\theta,I_r}t^{ppp,(Ef)}_{(Mi),(Nj),(Kl)}p_{Kl}\right] + \sum_{Mnk,ijl}^{I_\theta,I_r}t^{prr,(Ef)}_{(Mi),(nj),(kl)}\dot{p}_{Mi}\dot{r}_{nj}r_{kl}$$
$$+\sum_{mn,ij}^{I_\theta,I_r}\dot{r}_{mi}\dot{r}_{nj}\left[t^{rr,(Ef)}_{(mi),(nj)} + \sum_{K,l}^{I_\theta,I_r}t^{rrp,(Ef)}_{(mi),(nj),(Kl)}p_{Kl}\right] + \bar{\sigma}^2_{Ef}p_{Ef}$$
$$= -(\ddot{\eta}_1 - g\eta_5 - S_b\ddot{\eta}_5)\delta_{1E}\kappa_{1f}P_f; \quad E = 0,\ldots,I_\theta; \; f = 1,\ldots,I_r, \quad (4.30a)$$

$$\sum_{m,i}^{I_\theta,I_r}\ddot{r}_{mi}\left[\delta_{me}\delta_{ij} + \sum_{N,j}^{I_\theta,I_r}d^{rp,(ef)}_{(mi),(Nj)}p_{Nj} + \sum_{NK,jl}^{I_\theta,I_r}d^{rpp,(ef)}_{(mi),(Nj),(Kl)}p_{Nj}p_{Kl}\right.$$

$$+ \sum_{nk,jl}^{I_\theta, I_r} d^{rrr,(ef)}_{(mi),(nj),(kl)} r_{nj} r_{kl} \bigg] + \sum_{Mn,ij}^{I_\theta, I_r} \ddot{p}_{Mi} r_{nj} \bigg[d^{pr,(ef)}_{(Mi),(nj)} + \sum_{K,l}^{I_\theta, I_r} d^{prp,(ef)}_{(Mi),(nj),(Kl)} p_{Kl} \bigg]$$

$$+ \sum_{Mn,ij}^{I_\theta, I_r} \dot{p}_{Mi} \dot{r}_{nj} \bigg[t^{pr,(ef)}_{(Mi),(nj)} + \sum_{K,l}^{I_\theta, I_r} t^{prp,(ef)}_{(Mi),(nj),(Kl)} p_{Kl} \bigg]$$

$$+ \sum_{MNk,ijl}^{I_\theta, I_r} t^{ppr,(ef)}_{(Mi),(Nj),(kl)} \dot{p}_{Mi} \dot{p}_{Nj} r_{kl} + \sum_{mnk,ijl}^{I_\theta, I_r} t^{rrr,(ef)}_{(mi),(nj),(kl)} \dot{r}_{mi} \dot{r}_{nj} r_{kl} + \bar{\sigma}^2_{ef} r_{ef}$$

$$= -(\ddot{\eta}_2 + g\eta_4 + S_b \ddot{\eta}_4) \delta_{1e} \kappa_{1f} P_f; \quad e = 1, \ldots, I_\theta; \quad f = 1, \ldots, I_r, \quad (4.30b)$$

where the nondimensional natural sloshing frequencies $\bar{\sigma}_{Ef} = \sigma_{Et}/\sigma$ are defined by (4.7), P_f and S_b are given in (4.10) and (4.15), respectively, and the hydrodynamic coefficients of the nonlinear quantities can be explicitly calculated as functions of the nondimensional liquid depth h by the formulas

$$d^{pp,(Ef)}_{(Mi),(Nj)} = \frac{\kappa_{Ef}}{\Lambda_{EE}} \bigg[\Lambda_{EE} V^{pp}_{(Mi),(Nj),(Ef)} + \frac{\chi^{pp}_{(Nj)(Ef),(Mi)}}{\kappa_{Mi}} \bigg],$$

$$d^{ppp,(Ef)}_{(Mi),(Nj),(Kl)} = \frac{\kappa_{Ef}}{\Lambda_{EE}} \bigg[\Lambda_{EE} V^{ppp}_{(Mi),(Nj),(Kl),(Ef)} + \frac{\chi^{ppp}_{(Ef)(Nj)(Kl),(Mi)}}{\kappa_{Mi}}$$

$$+ \sum_{A,b}^{I_\theta, I_r} V^{pp}_{(Mi),(Nj),(Ab)} \chi^{pp}_{(Kl)(Ef),(Ab)} \bigg],$$

$$d^{prr,(Ef)}_{(Mi),(nj),(kl)} = \frac{\kappa_{Ef}}{\Lambda_{EE}} \bigg[\Lambda_{EE} V^{prr}_{(Mi),(nj),(kl),(Ef)} + \frac{\chi^{prr}_{(Ef),(nj)(kl),(Mi)}}{\kappa_{Mi}}$$

$$+ \sum_{a,b}^{I_\theta, I_r} V^{pr}_{(Mi),(nj),(ab)} \chi^{pr}_{(Ef),(kl),(ab)} \bigg],$$

$$d^{rr,(Ef)}_{(mi),(nj)} = \frac{\kappa_{Ef}}{\Lambda_{EE}} \bigg[\Lambda_{EE} V^{rr}_{(mi),(nj),(Ef)} + \frac{\chi^{pr}_{(Ef),(nj),(mi)}}{\kappa_{mi}} \bigg],$$

$$d^{rrp,(Ef)}_{(mi),(nj),(Kl)} = \frac{\kappa_{Ef}}{\Lambda_{EE}} \bigg[\Lambda_{EE} V^{rpr}_{(mi),(Kl),(nj),(Ef)} + \frac{2\chi^{ppr}_{(Kl)(Ef),(nj),(mi)}}{\kappa_{mi}}$$

$$+ \sum_{a,b}^{I_\theta, I_r} V^{rp}_{(mi),(Kl),(ab)} \chi^{pr}_{(Ef),(nj),(ab)} + \sum_{A,b}^{I_\theta, I_r} V^{rr}_{(mi),(nj),(Ab)} \chi^{pp}_{(Kl)(Ef),(Ab)} \bigg],$$

$$t^{pp,(Ef)}_{(Mi),(Nj)} = \frac{\kappa_{Ef}}{\Lambda_{EE}} \bigg[\Lambda_{EE} V^{pp}_{(Mi),(Nj),(Ef)} + \frac{\Pi^{p,p}_{(Ef),(Mi)(Nj)}}{2\kappa_{Mi}\kappa_{Nj}} \bigg],$$

$$t^{ppp,(Ef)}_{(Mi),(Nj),(Kl)} = \frac{\kappa_{Ef}}{\Lambda_{EE}} \left[\Lambda_{EE} \bar{V}^{ppp}_{(Mi),(Nj),(Kl),(Ef)} + \frac{\Pi^{p,pp}_{(Kl)(Ef),(Mi)(Nj)}}{\kappa_{Mi}\kappa_{Nj}} \right.$$
$$\left. + \sum_{A,b}^{I_\theta,I_r} V^{pp}_{(Mi),(Nj),(Ab)} \chi^{pp}_{(Kl)(Ef),(Ab)} + \sum_{A,b}^{I_\theta,I_r} \frac{\Pi^{p,p}_{(Ef),(Mi)(Ab)}}{\kappa_{Mi}} V^{pp}_{(Nj),(Kl),(Ab)} \right],$$

$$t^{rr,(Ef)}_{(mi),(nj)} = \frac{\kappa_{Ef}}{\Lambda_{EE}} \left[\Lambda_{EE} V^{rr}_{(mi),(nj),(Ef)} + \frac{\Pi^{r,p}_{(Ef),(mi)(nj)}}{2\kappa_{mi}\kappa_{nj}} \right],$$

$$t^{rrp,(Ef)}_{(mi),(nj),(Kl)} = \frac{\kappa_{Ef}}{\Lambda_{EE}} \left[\Lambda_{EE} V^{rpr}_{(mi),(Kl),(nj),(Ef)} + \frac{\Pi^{r,pp}_{(Kl)(Ef),(mi)(nj)}}{\kappa_{mi}\kappa_{nj}} \right.$$
$$\left. + \sum_{A,b}^{I_\theta,I_r} V^{rr}_{(mi),(nj),(Ab)} \chi^{pp}_{(Kl)(Ef),(Ab)} + \sum_{a,b}^{I_\theta,I_r} \frac{\Pi^{r,p}_{(Ef),(mi)(ab)}}{\kappa_{mi}} V^{rp}_{(nj),(Kl),(ab)} \right],$$

$$t^{prr,(Ef)}_{(Mi),(nj),(kl)} = \frac{\kappa_{Ef}}{\Lambda_{EE}} \left[\Lambda_{EE} \bar{V}^{prr}_{(Mi),(nj),(kl),(Ef)} + \frac{\Pi^{pr}_{(Ef),(kl),(Mi)(nj)}}{\kappa_{Mi}\kappa_{nj}} \right.$$
$$+ \sum_{a,b}^{I_\theta,I_r} \left(\bar{V}^{pr}_{(Mi),(nj),(ab)} \chi^{pr}_{(Ef),(kl),(ab)} + \frac{1}{\kappa_{nj}} V^{pr}_{(Mi),(kl),(ab)} \Pi^{r,p}_{(Ef),(ab)(nj)} \right)$$
$$\left. + \sum_{A,b}^{I_\theta,I_r} \frac{\Pi^{p,p}_{(Ef),(Mi)(Ab)}}{\kappa_{Mi}} V^{rr}_{(nj),(kl),(Ab)} \right],$$

$$d^{pr,(ef)}_{(Mi),(nj)} = \frac{\kappa_{ef}}{\Lambda_{ee}} \left[\Lambda_{ee} V^{pr}_{(Mi),(nj),(ef)} + \frac{\chi^{rr}_{(nj),(ef),(Mi)}}{\kappa_{Mi}} \right],$$

$$d^{prp,(ef)}_{(Mi),(nj),(Kl)} = \frac{\kappa_{ef}}{\Lambda_{ee}} \left[\Lambda_{ee} V^{ppr}_{(Mi),(Kl),(nj),(ef)} + \frac{2\chi^{prr}_{(Kl),(nj)(ef),(Mi)}}{\kappa_{Mi}} \right.$$
$$\left. + \sum_{A,b}^{I_\theta,I_r} V^{pp}_{(Mi),(Kl),(Ab)} \chi^{rr}_{(nj)(ef),(Ab)} + \sum_{a,b}^{I_\theta,I_r} V^{pr}_{(Mi),(nj),(ab)} \chi^{pr}_{(Kl),(ef),(ab)} \right],$$

$$d^{rp,(ef)}_{(mi),(Nj)} = \frac{\kappa_{ef}}{\Lambda_{ee}} \left[\Lambda_{ee} V^{rp}_{(mi),(Nj),(ef)} + \frac{\chi^{pr}_{(Nj),(ef),(mi)}}{\kappa_{mi}} \right],$$

$$d^{rpp,(ef)}_{(mi),(Nj),(Kl)} = \frac{\kappa_{ef}}{\Lambda_{ee}} \left[\Lambda_{ee} V^{rpp}_{(mi),(Nj),(Kl),(ef)} + \frac{\chi^{ppr}_{(Nj)(Kl),(ef),(mi)}}{\kappa_{mi}} \right.$$
$$\left. + \sum_{a,b}^{I_\theta,I_r} V^{rp}_{(mi),(Nj),(ab)} \chi^{pr}_{(Kl),(ef),(ab)} \right],$$

$$d^{rrr,(ef)}_{(mi),(nj),(kl)} = \frac{\kappa_{ef}}{\Lambda_{ee}} \left[\Lambda_{ee} V^{rrr}_{(mi),(nj),(kl),(ef)} + \frac{\chi^{rrr}_{(nj)(kl)(ef),(mi)}}{\kappa_{mi}} \right.$$

$$\left. + \sum_{A,b}^{I_\theta,I_r} V^{rr}_{(mi),(nj),(Ab)} \chi^{rr}_{(kl)(ef),(Ab)} \right],$$

$$t^{pr,(ef)}_{(Mi),(nj)} = \frac{\kappa_{ef}}{\Lambda_{ee}} \left[\Lambda_{ee} \bar{V}^{pr}_{(Mi),(nj),(ef)} + \frac{\Pi^{r}_{(ef),(Mi),(nj)}}{\kappa_{Mi}\kappa_{nj}} \right],$$

$$t^{prp,(ef)}_{(Mi),(nj),(Kl)} = \frac{\kappa_{ef}}{\Lambda_{ee}} \left[\Lambda_{ee} \bar{V}^{rpp}_{(nj),(Mi),(Kl),(ef)} + \frac{\Pi^{pr}_{(Kl),(ef),(Mi)(nj)}}{\kappa_{Mi}\kappa_{nj}} \right.$$

$$+ \sum_{a,b}^{I_\theta,I_r} \bar{V}^{pr}_{(Mi),(nj),(ab)} \chi^{pr}_{(Kl),(ef),(ab)} + \sum_{a,b}^{I_\theta,I_r} \frac{V^{rp}_{(nj),(Kl),(ab)}}{\kappa_{Mi}} \Pi^{r}_{(ef),(Mi),(ab)}$$

$$\left. + \sum_{A,b}^{I_\theta,I_r} \frac{V^{pp}_{(Mi),(Kl),(Ab)}}{\kappa_{nj}} \Pi^{r}_{(ef),(Ab),(nj)} \right],$$

$$t^{ppr,(ef)}_{(Mi),(Nj),(kl)} = \frac{\kappa_{ef}}{\Lambda_{ee}} \left[\Lambda_{ee} V^{ppr}_{(Mi),(Nj),(kl),(ef)} + \frac{\Pi^{p,rr}_{(kl)(ef),(Mi)(Nj)}}{\kappa_{Mi}\kappa_{Nj}} \right.$$

$$\left. + \sum_{A,b}^{I_\theta,I_r} V^{pp}_{(Mi),(Nj),(Ab)} \chi^{rr}_{(kl)(ef),(Ab)} + \sum_{a,b}^{I_\theta,I_r} \frac{V^{pr}_{(Nj),(kl),(ab)}}{\kappa_{Mi}} \Pi^{r}_{(ef),(Mi),(ab)} \right],$$

$$t^{rrr,(ef)}_{(mi),(nj),(kl)} = \frac{\kappa_{ef}}{\Lambda_{ee}} \left[\Lambda_{ee} \bar{V}^{rrr}_{(mi),(nj),(kl),(ef)} + \frac{\Pi^{r,rr}_{(kl)(ef),(mi)(nj)}}{\kappa_{mi}\kappa_{nj}} \right.$$

$$\left. + \sum_{A,b}^{I_\theta,I_r} V^{rr}_{(mi),(nj),(Ab)} \chi^{rr}_{(kl)(ef),(Ab)} + \sum_{A,b}^{I_\theta,I_r} \frac{V^{rr}_{(mi),(kl),(Ab)}}{\kappa_{nj}} \Pi^{r}_{(ef),(Ab),(nj)} \right],$$

where

$$\bar{V}^{ppp}_{(Mi),(Nj),(Kl),(Ab)} = V^{ppp}_{(Mi),(Nj),(Kl),(Ab)} + V^{ppp}_{(Mi),(Kl),(Nj),(Ab)},$$

$$\bar{V}^{prr}_{(Mi),(nj),(kl),(Ab)} = V^{prr}_{(Mi),(nj),(kl),(Ab)} + V^{prr}_{(Mi),(kl),(nj),(Ab)} + V^{rpr}_{(nj),(Mi),(kl),(Ab)},$$

$$\bar{V}^{pr}_{(Mi),(nj),(ab)} = V^{pr}_{(Mi),(nj),(ab)} + V^{rp}_{(nj),(Mi),(ab)},$$

$$\bar{V}^{rrr}_{(mi),(nj),(kl),(ab)} = V^{rrr}_{(mi),(nj),(kl),(ab)} + V^{rrr}_{(mi),(kl),(nj),(ab)},$$

$$\bar{V}^{rpp}_{(kl),(Mi),(Nj),(ab)} = V^{rpp}_{(kl),(Mi),(Nj),(ab)} + V^{rpp}_{(kl),(Nj),(Mi),(ab)} + V^{ppr}_{(Mi),(Nj),(kl),(ab)}.$$

The system of ordinary differential equations (4.30) can be interpreted as a discrete conservative mechanical system with an infinite number of degrees of freedom. Neglecting the nonlinear quantities leads to the linear modal equations, which are based on normalisation (4.4) and (4.6) instead of (3.56).

70   *Sloshing in Upright Circular Containers*

Adopting the generic adaptive nonlinear modal equations (4.30) for simulating the transient sloshing, or, in other words, conducting the Perko-type simulation is, as discussed in the Introduction (Chapter 1), a rather disputable way. We already mentioned that these time-step integrations are very stiff and, normally, lead to an unrealistic amplification of higher harmonics. One should 'filter' these harmonics either numerically or analytically, excluding physically negligible higher-order hydrodynamic generalised coordinates. This can, for instance, be done by adopting the Narimanov-Moiseev asymptotic approximation.

4.3 NARIMANOV-MOISEEV–TYPE MODAL THEORY

4.3.1 NARIMANOV-MOISEEV–TYPE MODAL EQUATIONS

Chapter 1 introduces the Narimanov-Moiseev asymptotic modal approach to the nonlinear resonant sloshing. It assumes that the forcing frequency σ is close to the lowest natural sloshing frequency σ_{11} so that the Moiseev detuning (1.1) is satisfied and the mechanical system is not exposed to the secondary resonance (according to [52], this requires the nondimensional liquid depth $h \gtrsim 1.05$ for the circular base tank). The asymptotic approach requires special relations between the hydrodynamic generalised coordinates and velocities:

$$p_{11} \sim r_{11} = O(\epsilon^{1/3}), \quad p_{0j} \sim p_{2j} \sim r_{2j} = O(\epsilon^{2/3}),$$
$$r_{1(j+1)} \sim p_{1(j+1)} \sim p_{3j} \sim r_{3j} = O(\epsilon), \quad j = 1, 2, \ldots, I_r; \quad I_r \to \infty \quad (4.31a)$$

$$\Lambda = \bar{\sigma}_{11}^2 - 1 = O(\epsilon^{2/3}), \quad \bar{\sigma}_{Mi} = \sigma_{Mi}/\sigma. \quad (4.31b)$$

Accounting for (4.31a) in the adaptive modal equations (4.30) and neglecting the $o(\epsilon)$-terms derives the following nonlinear modal equations of the Narimanov-Moiseev type with respect to the hydrodynamic generalised coordinates

$$\ddot{p}_{11} + \boxed{2\xi_{11}\bar{\sigma}_{11}\dot{p}_{11}} + \bar{\sigma}_{11}^2 p_{11} + \mathcal{P}_{11}(p_{11}, r_{11}; p_{0j}, p_{2j}, r_{2j}) = -(\ddot{\eta}_1 - g\eta_5 - S_1\ddot{\eta}_5)\kappa_{11}P_1, \quad (4.32a)$$

$$\ddot{r}_{11} + \boxed{2\xi_{11}\bar{\sigma}_{11}\dot{r}_{11}} + \bar{\sigma}_{11}^2 r_{11} + \mathcal{Q}_{11}(p_{11}, r_{11}; p_{0j}, p_{2j}, r_{2j}) = -(\ddot{\eta}_2 + g\eta_4 + S_1\ddot{\eta}_4)\kappa_{11}P_1; \quad (4.32b)$$

$$\ddot{p}_{2k} + \boxed{2\xi_{2k}\bar{\sigma}_{2k}\dot{p}_{2k}} + \bar{\sigma}_{2k}^2 p_{2k} + \mathcal{P}_{2k}(p_{11}, r_{11}) = 0, \quad (4.33a)$$

$$\ddot{r}_{2k} + \boxed{2\xi_{2k}\bar{\sigma}_{2k}\dot{r}_{2k}} + \bar{\sigma}_{2k}^2 r_{2k} + \mathcal{Q}_{2k}(p_{11}, r_{11}) = 0, \quad (4.33b)$$

$$\ddot{p}_{0k} + \boxed{2\xi_{0k}\bar{\sigma}_{0k}\dot{p}_{0k}} + \bar{\sigma}_{0k}^2 p_{0k} + \mathcal{P}_{0k}(p_{11}, r_{11}) = 0; \quad (4.33c)$$

$$\ddot{p}_{3k} + \boxed{2\xi_{3k}\bar{\sigma}_{3k}\dot{p}_{3k}} + \bar{\sigma}_{3k}^2 p_{3k} + \mathcal{P}_{3k}(p_{11}, r_{11}; p_{0j}, p_{2j}, r_{2j}) = 0, \quad (4.34a)$$

$$\ddot{r}_{3k} + \boxed{2\xi_{3k}\bar{\sigma}_{3k}\dot{r}_{3k}} + \bar{\sigma}_{3k}^2 r_{3k} + \mathcal{Q}_{3k}(p_{11}, r_{11}; p_{0j}, p_{2j}, r_{2j}) = 0, \ k = 1, ..., I_r; \quad (4.34b)$$

$$\ddot{p}_{1n} + \boxed{2\xi_{1n}\bar{\sigma}_{1n}\dot{p}_{1n}} + \bar{\sigma}_{1n}^2 p_{1n} + \mathcal{P}_{1n}(p_{11}, r_{11}; p_{0j}, p_{2j}, r_{2j}) =$$
$$= -(\ddot{\eta}_1 - g\eta_5 - S_n \ddot{\eta}_5)\kappa_{1n}P_n, \quad (4.35a)$$

$$\ddot{r}_{1n} + \boxed{2\xi_{1n}\bar{\sigma}_{1n}\dot{r}_{1n}} + \bar{\sigma}_{1n}^2 r_{1n} + \mathcal{Q}_{1n}(p_{11}, r_{11}; p_{0j}, p_{2j}, r_{2j}) =$$
$$= -(\ddot{\eta}_2 + g\eta_4 + S_n \ddot{\eta}_4)\kappa_{1n}P_n, \quad n = 2, ..., I_r, \quad (4.35b)$$

where P_n and S_n are computed by (4.10) and (4.15), respectively, κ_{Mi} is introduced in (4.7), but $\mathcal{P}_{Mi}, \mathcal{Q}_{mi}$ take the following form

$$\mathcal{P}_{11}(p_{11}, r_{11}; p_{0j}, p_{2j}, r_{2j}) = d_1 p_{11} \left(\ddot{p}_{11} p_{11} + \ddot{r}_{11} r_{11} + \dot{p}_{11}^2 + \dot{r}_{11}^2 \right)$$
$$+ d_2 \left[r_{11}(\ddot{p}_{11}r_{11} - \ddot{r}_{11}p_{11}) + 2\dot{r}_{11}(\dot{p}_{11}r_{11} - \dot{r}_{11}p_{11}) \right]$$
$$+ \sum_{j=1}^{I_r} \left[d_3^{(j)} (\ddot{p}_{11}p_{2j} + \ddot{r}_{11}r_{2j} + \dot{p}_{11}\dot{p}_{2j} + \dot{r}_{11}\dot{r}_{2j}) + d_4^{(j)} (\ddot{p}_{2j}p_{11} + \ddot{r}_{2j}r_{11}) \right.$$
$$\left. + d_5^{(j)} (\ddot{p}_{11}p_{0j} + \dot{p}_{11}\dot{p}_{0j}) + d_6^{(j)} \ddot{p}_{0j}p_{11} \right], \quad (4.36a)$$

$$\mathcal{Q}_{11}(p_{11}, r_{11}; p_{0j}, p_{2j}, r_{2j}) = d_1 r_{11} \left(\ddot{p}_{11} p_{11} + \ddot{r}_{11} r_{11} + \dot{p}_{11}^2 + \dot{r}_{11}^2 \right)$$
$$+ d_2 \left[p_{11}(\ddot{r}_{11}p_{11} - \ddot{p}_{11}r_{11}) + 2\dot{p}_{11}(\dot{r}_{11}p_{11} - \dot{p}_{11}r_{11}) \right]$$
$$+ \sum_{j=1}^{I_r} \left[d_3^{(j)} (\ddot{p}_{11}r_{2j} - \ddot{r}_{11}p_{2j} + \dot{p}_{11}\dot{r}_{2j} - \dot{p}_{2j}\dot{r}_{11}) + d_4^{(j)} (\ddot{r}_{2j}p_{11} - \ddot{p}_{2j}r_{11}) \right.$$
$$\left. + d_5^{(j)} (\ddot{r}_{11}p_{0j} + \dot{r}_{11}\dot{p}_{0j}) + d_6^{(j)} \ddot{p}_{0j}r_{11} \right], \quad (4.36b)$$

$$\mathcal{P}_{2k}(p_{11}, r_{11}) = d_{7,k}(\dot{p}_{11}^2 - \dot{r}_{11}^2) + d_{9,k}(\ddot{p}_{11}p_{11} - \ddot{r}_{11}r_{11}), \quad (4.37a)$$

$$\mathcal{Q}_{2k}(p_{11}, r_{11}) = 2d_{7,k}\dot{p}_{11}\dot{r}_{11} + d_{9,k}(\ddot{p}_{11}r_{11} + \ddot{r}_{11}p_{11}) = 0, \quad (4.37b)$$

$$\mathcal{P}_{0k}(p_{11}, r_{11}) = d_{8,k}(\dot{p}_{11}^2 + \dot{r}_{11}^2) + d_{10,k}(\ddot{p}_{11}p_{11} + \ddot{r}_{11}r_{11}); \quad (4.37c)$$

$$\mathcal{P}_{3k}(p_{11}, r_{11}; p_{0j}, p_{2j}, r_{2j}) = d_{11,k} \left[\ddot{p}_{11}(p_{11}^2 - r_{11}^2) - 2p_{11}r_{11}\ddot{r}_{11} \right]$$
$$+ d_{12,k} \left[p_{11}(\dot{p}_{11}^2 - \dot{r}_{11}^2) - 2r_{11}\dot{p}_{11}\dot{r}_{11} \right] + \sum_{j=1}^{I_r} \left[d_{13,k}^{(j)} (\ddot{p}_{11}p_{2j} - \ddot{r}_{11}r_{2j}) \right.$$
$$\left. + d_{14,k}^{(j)} (\ddot{p}_{2j}p_{11} - \ddot{r}_{2j}r_{11}) + d_{15,k}^{(j)} (\dot{p}_{2j}\dot{p}_{11} - \dot{r}_{2j}\dot{r}_{11}) \right], \quad (4.38a)$$

$$\mathcal{Q}_{3k}(p_{11}, r_{11}; p_{0j}, p_{2j}, r_{2j}) = d_{11,k} \left[\ddot{r}_{11}(p_{11}^2 - r_{11}^2) + 2p_{11}r_{11}\ddot{p}_{11} \right]$$

$$+ d_{12,k} \left[r_{11}(\dot{p}_{11}^2 - \dot{r}_{11}^2) + 2 p_{11}\dot{p}_{11}\dot{r}_{11} \right] + \sum_{j=1}^{I_r} \left[d_{13,k}^{(j)}(\ddot{p}_{11}r_{2j} + \ddot{r}_{11}p_{2j}) \right.$$

$$\left. + d_{14,k}^{(j)}(\ddot{p}_{2j}r_{11} + \ddot{r}_{2j}p_{11}) + d_{15,k}^{(j)}(\dot{p}_{2j}\dot{r}_{11} + \dot{r}_{2j}\dot{p}_{11}) \right]; \quad (4.38b)$$

$$\mathcal{P}_{1n}(p_{11}, r_{11}; p_{0j}, p_{2j}, r_{2j}) = d_{16,n}(\ddot{p}_{11}p_{11}^2 + r_{11}p_{11}\ddot{r}_{11})$$
$$+ d_{17,n}(\ddot{p}_{11}r_{11}^2 - r_{11}p_{11}\ddot{r}_{11}) + d_{18,n}p_{11}(\dot{p}_{11}^2 + \dot{r}_{11}^2) + d_{19,n}(r_{11}\dot{p}_{11}\dot{r}_{11} - p_{11}\dot{r}_{11}^2)$$
$$+ \sum_{j=1}^{I_r} \left[d_{20,n}^{(j)}(\ddot{p}_{11}p_{2j} + \ddot{r}_{11}r_{2j}) + d_{21,n}^{(j)}(p_{11}\ddot{p}_{2j} + r_{11}\ddot{r}_{2j}) \right.$$
$$\left. + d_{22,n}^{(j)}(\dot{p}_{11}\dot{p}_{2j} + \dot{r}_{11}\dot{r}_{2j}) + d_{23,n}^{(j)}\ddot{p}_{11}p_{0j} + d_{24,n}^{(j)}p_{11}\ddot{p}_{0j} + d_{25,n}^{(j)}\dot{p}_{11}\dot{p}_{0j} \right], \quad (4.39a)$$

$$\mathcal{Q}_{1n}(p_{11}, r_{11}; p_{0j}, p_{2j}, r_{2j}) = d_{16,n}(\ddot{r}_{11}r_{11}^2 + r_{11}p_{11}\ddot{p}_{11})$$
$$+ d_{17,n}(\ddot{r}_{11}p_{11}^2 - r_{11}p_{11}\ddot{p}_{11}) + d_{18,n}r_{11}(\dot{p}_{11}^2 + \dot{r}_{11}^2) + d_{19,n}(p_{11}\dot{p}_{11}\dot{r}_{11} - r_{11}\dot{p}_{11}^2)$$
$$+ \sum_{j=1}^{I_r} \left[d_{20,n}^{(j)}(\ddot{p}_{11}r_{2j} - \ddot{r}_{11}p_{2j}) + d_{21,n}^{(j)}(p_{11}\ddot{r}_{2j} - r_{11}\ddot{p}_{2j}) + d_{22,n}^{(j)}(\dot{p}_{11}\dot{r}_{2j} - \dot{r}_{11}\dot{p}_{2j}) \right.$$
$$\left. + d_{23,n}^{(j)}\ddot{r}_{11}p_{0j} + d_{24,n}^{(j)}r_{11}\ddot{p}_{0j} + d_{25,n}^{(j)}\dot{r}_{11}\dot{p}_{0j} \right], \quad n = 2, ..., I_r. \quad (4.39b)$$

Here, the nondimensional hydrodynamic coefficients are functions of the nondimensional liquid depth h [52]. They are computed by the formulas

$$d_1 = d_{(11),(11),(11)}^{ppp,(11)} = d_{(11),(11),(11)}^{rrr,(11)} = t_{(11),(11),(11)}^{ppp,(11)} = t_{(11),(11),(11)}^{rrr,(11)},$$

$$d_2 = d_{(11),(11),(11)}^{prr,(11)} = d_{(11),(11),(11)}^{rpp,(11)} = \tfrac{1}{2} t_{(11),(11),(11)}^{prr,(11)} = \tfrac{1}{2} t_{(11),(11),(11)}^{prp,(11)},$$

$$d_1 - d_2 = d_{(11),(11),(11)}^{rrp,(11)} = d_{(11),(11),(11)}^{prp,(11)}, \quad d_1 - 2d_2 = t_{(11),(11),(11)}^{rrp,(11)} = t_{(11),(11),(11)}^{ppr,(11)},$$

$$d_3^{(j)} = d_{(11),(2j)}^{pp,(11)} = d_{(11),(2j)}^{rr,(11)} = d_{(11),(2j)}^{pr,(11)} = -d_{(11),(2j)}^{rp,(11)} = t_{(11),(2j)}^{pr,(11)} = -t_{(2j),(11)}^{pr,(11)}$$
$$= t_{(11),(2j)}^{pp,(11)} + t_{(2j),(11)}^{pp,(11)} = t_{(11),(2j)}^{rr,(11)} + t_{(2j),(11)}^{rr,(11)},$$

$$d_4^{(j)} = d_{(2j),(11)}^{pp,(11)} = d_{(2j),(11)}^{rr,(11)} = -d_{(2j),(11)}^{pr,(11)} = d_{(2j),(11)}^{rp,(11)},$$

$$d_5^{(j)} = d_{(11),(0j)}^{pp,(11)} = d_{(11),(0j)}^{rp,(11)} = t_{(0j),(11)}^{pr,(11)} = t_{(0j),(11)}^{pp,(11)} + t_{(11),(0j)}^{pp,(11)},$$

$$d_6^{(j)} = d_{(0j),(11)}^{pp,(11)} = d_{(0j),(11)}^{pr,(11)}, \quad d_{7,k} = t_{(11),(11)}^{pp,(2k)} = -t_{(11),(11)}^{rr,(2k)} = \tfrac{1}{2} t_{(11),(11)}^{pr,(2k)},$$

$$d_{8,k} = t_{(11),(11)}^{pp,(0k)} = t_{(11),(11)}^{rr,(0k)}, \quad d_{10,k} = d_{(11),(11)}^{pp,(0k)} = d_{(11),(11)}^{rr,(0k)},$$

$$d_{9,k} = d_{(11),(11)}^{pp,(2k)} = -d_{(11),(11)}^{rr,(2k)} = d_{(11),(11)}^{pr,(2k)} = d_{(11),(11)}^{rp,(2k)},$$

$$d_{11,k} = d_{(11),(11),(11)}^{ppp,(3k)} = -d_{(11),(11),(11)}^{rrr,(3k)} = -\tfrac{1}{2} d_{(11),(11),(11)}^{rrp,(3k)},$$

$$d_{12,k} = t_{(11),(11),(11)}^{ppp,(3k)} = -t_{(11),(11),(11)}^{rrp,(3k)} = -\tfrac{1}{2} t_{(11),(11),(11)}^{prr,(3k)} = t_{(11),(11),(11)}^{ppr,(3k)}$$

$$= -t^{rrr,(3k)}_{(11),(11),(11)} = \tfrac{1}{2} t^{prp,(3k)}_{(11),(11),(11)},$$

$$d^{(j)}_{13,k} = d^{pp,(3k)}_{(11),(2j)} = -d^{rr,(3k)}_{(11),(2j)} = d^{pr,(3k)}_{(11),(2j)} = d^{rp,(3k)}_{(11),(2j)},$$

$$d^{(j)}_{14,k} = d^{pp,(3k)}_{(2j),(11)} = -d^{rr,(3k)}_{(2j),(11)} = d^{pr,(3k)}_{(2j),(11)} = d^{rp,(3k)}_{(2j),(11)},$$

$$d^{(j)}_{15,k} = t^{pp,(3k)}_{(11),(2j)} + t^{pp,(3k)}_{(2j),(11)} = -t^{rr,(3k)}_{(11),(2j)} - t^{rr,(3k)}_{(2j),(11)} = t^{pr,(3k)}_{(11),(2j)} = t^{pr,(3k)}_{(2j),(11)},$$

$$d_{16,n} = d^{ppp,(1n)}_{(11),(11),(11)} = d^{rrr,(1n)}_{(11),(11),(11)}, \quad d_{17,n} = d^{prr,(1n)}_{(11),(11),(11)} = d^{rpp,(1n)}_{(11),(11),(11)},$$

$$d_{18,n} = t^{ppp,(1n)}_{(11),(11),(11)} = t^{rrr,(1n)}_{(11),(11),(11)}, \quad d_{19,n} = t^{prr,(1n)}_{(11),(11),(11)} = t^{prp,(1n)}_{(11),(11),(11)},$$

$$d_{16,n} - d_{17,n} = d^{rrp,(1n)}_{(11),(11),(11)} = d^{prp,(1n)}_{(11),(11),(11)},$$

$$d_{18,n} - d_{19,n} = t^{rrp,(1n)}_{(11),(11),(11)} = t^{ppr,(1n)}_{(11),(11),(11)},$$

$$d^{(j)}_{20,n} = d^{pp,(1n)}_{(11),(2j)} = d^{rr,(1n)}_{(11),(2j)} = d^{pr,(1n)}_{(11),(2j)} = -d^{rp,(1n)}_{(11),(2j)},$$

$$d^{(j)}_{21,n} = d^{pp,(1n)}_{(2j),(11)} = d^{rr,(1n)}_{(2j),(11)} = -d^{pr,(1n)}_{(2j),(11)} = d^{rp,(1n)}_{(2j),(11)},$$

$$d^{(j)}_{22,n} = t^{pp,(1n)}_{(11),(2j)} + t^{pp,(1n)}_{(2j),(11)} = t^{rr,(1n)}_{(11),(2j)} + t^{rr,(1n)}_{(2j),(11)} = t^{pr,(1n)}_{(11),(2j)} = -t^{pr,(1n)}_{(2j),(11)},$$

$$d^{(j)}_{23,n} = d^{pp,(1n)}_{(11),(0j)} = d^{rp,(1n)}_{(11),(0j)}; \quad d^{(j)}_{24,n} = d^{pp,(1n)}_{(0j),(11)} = d^{pr,(1n)}_{(0j),(11)},$$

$$d^{(j)}_{25,n} = t^{pp,(1n)}_{(11),(0j)} + t^{pp,(1n)}_{(0j),(11)} = t^{pr,(1n)}_{(0j),(11)}.$$

Coefficients P_i, S_i, κ_{1i} and the nondimensional $\bar{\sigma}^2_{Mi}$ are explicitly computed by (4.10), (4.15), and (4.6) by using the roots of $J'_M(k_{Mi}) = 0$ and values of $J_M(k_{Mi})$ (they are listed in Tables 8.1 and 8.2). The nondimensional hydrodynamic coefficient at the nonlinear terms are complex functions of h. To motivate interested engineers to employ the Narimanov-Moiseev–type modal equations in analytical studies and simulations, we provide values of these coefficients in Chapter 8.

4.3.2 COMPARISON WITH EXISTING MODAL EQUATIONS

The Narimanov-Moiseev–type modal equations and associated hydrodynamic coefficients at the nonlinear terms can be compared with those in [143, 147] by Lukovsky (1990). The latter equations have only five degrees of freedom and use the following approximate solution

$$\zeta(r,z,\theta) = \frac{\mathcal{R}_{01}(r)}{\mathcal{R}_{01}(1)} p_0(t) + \frac{\mathcal{R}_{11}(r)}{\mathcal{R}_{11}(1)} \left[p_1(t) \cos\theta + r_1(t) \sin\theta \right]$$
$$+ \frac{\mathcal{R}_{21}(r)}{\mathcal{R}_{21}(1)} \left[p_2(t) \cos 2\theta + r_2(t) \sin 2\theta \right]. \quad (4.40)$$

This approximation completely neglects the third-order contribution and excludes the second-order generalised coordinates with the indices $(0n), (2n), n \geq 2$. The five-dimensional modal system by Lukovsky (1990) reads as [147, Eqs. (4.1.15)–(4.1.19)]

$$\mu_1(\ddot{r}_1 + \sigma_1^2 r_1) + d_1(r_1^2\ddot{r}_1 + r_1\dot{r}_1^2 + r_1 p_1\ddot{p}_1 + r_1\dot{p}_1^2) + d_2(p_1^2\ddot{r}_1 + 2p_1\dot{r}_1\dot{p}_1$$
$$- r_1 p_1\ddot{p}_1 - 2r_1\dot{p}_1^2) - d_3(r_2\ddot{r}_1 - r_2\ddot{p}_1 + \dot{r}_1\dot{p}_2 - \dot{p}_1\dot{r}_2) + d_4(r_1\ddot{p}_2 - p_1\ddot{r}_2)$$
$$+ d_5(p_0\ddot{r}_1 + \dot{r}_1\dot{p}_0) + d_6 r_1\ddot{p}_0 = -P\ddot{\eta}_2(t), \quad (4.41\text{a})$$

$$\mu_1(\ddot{p}_1 + \sigma_1^2 p_1) + d_1(p_1^2\ddot{p}_1 + r_1 p_1\ddot{r}_1 + p_1\dot{r}_1^2 + p_1\dot{p}_1^2) + d_2(r_1^2\ddot{p}_1 - r_1 p_1\ddot{r}_1$$
$$+ 2r_1\dot{r}_1\dot{p}_1 - 2p_1\dot{r}_1^2) + d_3(p_2\ddot{p}_1 + r_2\ddot{r}_1 + \dot{r}_1\dot{r}_2 + \dot{p}_1\dot{p}_2) - d_4(p_1\ddot{p}_2 + r_1\ddot{r}_2)$$
$$+ d_5(p_0\ddot{p}_1 + \dot{p}_1\dot{p}_0) + d_6 p_1\ddot{p}_0 = -P\ddot{\eta}_1(t), \quad (4.41\text{b})$$

$$\mu_0(\ddot{p}_0 + \sigma_0^2 p_0) + d_6(r_1\ddot{r}_1 + p_1\ddot{p}_1) + d_8(\dot{r}_1^2 + \dot{p}_1^2) = 0, \quad (4.41\text{c})$$

$$\mu_2(\ddot{r}_2 + \sigma_2^2 r_2) - d_4(p_1\ddot{r}_1 + r_1\ddot{p}_1) - 2d_7\dot{r}_1\dot{p}_1 = 0, \quad (4.41\text{d})$$

$$\mu_2(\ddot{p}_2 + \sigma_2^2 p_2) + d_4(r_1\ddot{r}_1 - p_1\ddot{p}_1) + d_7(\dot{r}_1^2 - \dot{p}_1^2) = 0. \quad (4.41\text{e})$$

Because Lukovsky used normalization of the linear sloshing theory, the following formulas are needed to link d_i and d_i

$$d_1 = \frac{\mathrm{d}_1 \mathcal{R}_{11}^2(1)}{\mu_1}, \quad d_2 = \frac{\mathrm{d}_2 \mathcal{R}_{11}^2(1)}{\mu_1}, \quad d_3^{(1)} = \frac{\mathrm{d}_3 \mathcal{R}_{21}(1)}{\mu_1}, \quad d_4^{(1)} = -\frac{\mathrm{d}_4 \mathcal{R}_{21}(1)}{\mu_1},$$
$$d_5^{(1)} = \frac{\mathrm{d}_5 \mathcal{R}_{01}(1)}{\mu_1}, \quad d_6^{(1)} = \frac{\mathrm{d}_6 \mathcal{R}_{01}(1)}{\mu_1}, \quad d_{7,1} = -\frac{\mathrm{d}_7 \mathcal{R}_{11}^2(1)}{\mathcal{R}_{21}(1)\mu_2}, \quad (4.42)$$
$$d_{8,1} = \frac{\mathrm{d}_8 \mathcal{R}_{11}^2(1)}{\mathcal{R}_{01}(1)\mu_0}, \quad d_{9,1} = -\frac{\mathrm{d}_4 \mathcal{R}_{11}^2(1)}{\mathcal{R}_{21}(1)\mu_2}, \quad d_{10,1} = \frac{\mathrm{d}_6 \mathcal{R}_{11}^2(1)}{\mathcal{R}_{01}(1)\mu_0}.$$

Our computations with $I_\theta = 2$ and $I_r = 1$ and the tabled hydrodynamic coefficients in [143, 147] (rescaled by (4.42)) have a small discrepancy, which is less than 0.1%, which is actually the announced precision in [143, 147]. This fact is illustrated in Figure 4.2. Increasing $I_r \geq 2$ increases the number of ordinary differential (modal) equations and may change the hydrodynamic coefficients at the cubic terms. One reason is that these hydrodynamic coefficients ($d_1, d_2, d_{11,k}, d_{12,k}, d_{16,k}$ and $d_{18,k}$) are functions of the upper summation limit I_r. This means, in particular, that d_1 and d_2 in (4.41) cannot be used in the Narimanov-Moiseev-type modal systems with $I_r \geq 2$.

4.3.3 SECONDARY RESONANCE

The nonlinear modal theory (4.32)–(4.35) assumes that the forcing amplitude is sufficiently small and the forcing frequency σ is close to the lowest natural sloshing frequency σ_{11}. Another limitation is that the theory requires no *secondary resonances*. Trigonometrical algebra by the angular coordinate θ determines the analytical structure of (4.33)–(4.35) where the dominant (lowest-order) generalised coordinates p_{11} and r_{11} (and their time-derivatives) constitute quadratic terms in the modal equations with respect to p_{2k}, r_{2k}, and p_{0k} (here, in (4.33)), but the cubic quantities in p_{11} and r_{11} also appear in

Nonlinear modal theories of nonparametric resonant waves

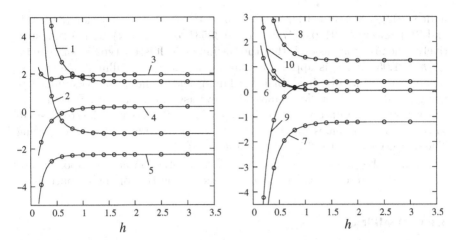

Figure 4.2 Hydrodynamic coefficients of the Narimanov–Moiseev–type modal system within the framework of the five-dimensional approximation ($I_\theta = 2, I_r = 1$) and computed in [143, 147] for different nondimensional liquid depths h. The solid lines denote our calculations, but the circles correspond to the tabled values from [143, 147] rescaled according to (4.42). The integers at the graphs imply: 1 – d_1, 2 – d_2, 3 – $d_3^{(1)}$, 4 – $d_4^{(1)}$, 5 – $d_5^{(1)}$, 6 – $d_6^{(1)}$, 7 – $d_{7,1}$, 8 – $d_{8,1}$, 9 – $d_{9,1}$, and 10 – $d_{10,1}$.

(4.32) and (4.34)–(4.35). When the forcing frequency σ is close to the lowest natural sloshing frequency σ_{11}, the generalised coordinates p_{11} and r_{11} contain the resonantly-excited harmonics $\cos t$ and $\sin t$ and quadratic terms by p_{11} and r_{11} in (4.33) yield the second Fourier harmonics. If these double harmonics are close to one of the natural sloshing frequencies $\{\bar{\sigma}_{2k}, \bar{\sigma}_{0k}\}$, a resonant amplification of the corresponding second-order generalised coordinate occurs. Analogously, cubic quantities in (4.34)–(4.35) yield the third Fourier harmonics, which can be close to $\bar{\sigma}_{3n}, n \geq 1$, and $\bar{\sigma}_{1k}, k \geq 2$, and thereby resonantly amplify the third-order generalised coordinates. Formalising this fact implies that the secondary resonance is expected when at least one of the following conditions

$$\sigma_{0k}/\sigma = 2, \quad \sigma_{2k}/\sigma = 2, \quad k \geq 1 \text{ (second-order resonances)} \quad (4.43\text{a})$$

$$\sigma_{1(k+1)}/\sigma = 3, \quad \sigma_{3k}/\sigma = 3, \quad k \geq 1 \text{ (third-order resonances)} \quad (4.43\text{b})$$

$$\sigma_{0k}/\sigma = 4, \quad \sigma_{2k}/\sigma = 4, \quad \sigma_{4k}/\sigma = 4, \quad k \geq 1 \text{ (fourth-order resonances)} \quad (4.43\text{c})$$

....

is satisfied accompanied by the primary resonance condition (4.31b).

Taking the asymptotic limit $\sigma = \sigma_{11}$ ($\Lambda = 0$ in (4.31b)) makes it possible to compute critical values of the mean liquid depths h from the equalities (4.43). The computations were done in [52] where the critical values were estimated at 1.0097, 0.83138, and 0.15227 for the generalised coordinates (modes) with indices (02), (22), and (01); at 0.74775, 0.45505, and 0.27897 for (34), (33), and (32); and at 0.79110, 0.47794, and 0.30531 for (15), (14), and (13), respectively. The first conclusion is that the Narimanov-Moiseev–type theory can fail for $h \lesssim 1.05$, i.e., it is applicable only for a fairly deep filling. The existing theoretical results obtained for lower liquid depths should be reexamined by using an adaptive (not Narimanov-Moiseev) asymptotic theory. This includes the well-known result on the *critical depth* $h = 0.5059$, where the steady-state wave response becomes infinite owing to the longitudinal harmonic forcing. The depth was discussed in [164] and [62, p. 425].

Because the modal theory is of the asymptotic character, importance of the secondary resonances increases with increasing the forcing amplitude.

4.3.4 DAMPING

A novelty of the derived asymptotic modal equations (4.32)–(4.35) consists of the framed linear damping terms, which are not considered in [52]. By definition, ξ_{Mi} should be small and we must keep only the $O(\epsilon^{2/3})$ and $O(\epsilon)$ quantities. This deduces that ξ_{Mi} should satisfy the asymptotic relation

$$\xi_{2i} \sim \xi_{0i} \sim \xi_{3i} \sim \xi_{1n} = O(1), \quad i \geq 1, n \geq 2, \qquad (4.44)$$

which looks non-physical and incompatible with our assumptions that the damping is small enough. Thus, we can omit the linear damping from the Narimanov-Moiseev–type modal system everywhere, except in equations (4.32), which deals with the primary resonance when even small damping may dramatically affect the behaviour of p_{11} and r_{11}.

4.4 HYDRODYNAMIC FORCE AND MOMENT

Lukovsky's formulas (2.71) and (2.80) facilitate computing the resulting hydrodynamic force and moment based on the modal solution. Our task consists of deriving particular Lukovsky's expressions for adaptive and Narimanov-Moiseev–type multimodal theories.

4.4.1 FORCE

Adopting the asymptotic modal solution (4.11a) and (2.71), as well as remembering Remarks 2.3 and 4.1, one can write down the adaptive approximation

$$\begin{aligned}\boldsymbol{F}(t) &= \left[M_l R_0 \sigma^2\right] \left(\boldsymbol{g} - \overset{*}{\boldsymbol{v}}_O - \overset{*}{\boldsymbol{\omega}} \times \boldsymbol{r}_{lC_0} - \overset{**}{\boldsymbol{r}}_{lC} + o(\epsilon)\right) \\ &= \left[M_l R_0 \sigma^2\right] \left(F_1(t)\,\boldsymbol{e}_1(t) + F_2(t)\,\boldsymbol{e}_2(t) + F_3(t)\,\boldsymbol{e}_3(t) + o(\epsilon)\right), \end{aligned} \qquad (4.45)$$

where the liquid mass centre (2.69) consists of hydrostatic $\boldsymbol{r}_{lC_0}$ and sloshing-related $\boldsymbol{r}_{lC_s} = r_{1lC_s}\boldsymbol{e}_1 + r_{2lC_s}\boldsymbol{e}_2 + r_{3lC_s}\boldsymbol{e}_3$ components,

$$\boldsymbol{r}_{lC} = \boldsymbol{r}_{lC_0} + \boldsymbol{r}_{lC_o} = \boldsymbol{e}_1 \left(\frac{1}{h} \sum_i^{I_r} P_i p_{1i}(t) \right) + \boldsymbol{e}_2 \left(\frac{1}{h} \sum_i^{I_r} P_i r_{1i}(t) \right)$$

$$+ \boldsymbol{e}_3 \left(-\frac{h}{2} + \frac{1}{h} \sum_{i=1}^{I_r} p_{0i}^2 + \frac{1}{2h} \sum_{m,i}^{I_\theta, I_r} (p_{mi}^2 + r_{mi}^2) \right), \quad (4.46)$$

where P_i are determined by (4.10). This deduces

$$F_1(t) = g\eta_5 - \ddot{\eta}_1 + \tfrac{1}{2} h \ddot{\eta}_5 - h^{-1} \sum_i^{I_r} P_i \ddot{p}_{1i}, \quad (4.47a)$$

$$F_2(t) = -g\eta_4 - \ddot{\eta}_2 - \tfrac{1}{2} h \ddot{\eta}_4 - h^{-1} \sum_i^{I_r} P_i \ddot{r}_{1i}, \quad (4.47b)$$

$$F_3(t) = -g - \ddot{\eta}_3 - 2h^{-1} \sum_i^{I_r} \left(\ddot{p}_{0i} p_{01} + \dot{p}_{0i}^2 \right)$$

$$- h^{-1} \sum_{m,i}^{I_\theta, I_r} \left(\ddot{p}_{mi} p_{mi} + \dot{p}_{mi}^2 + \ddot{r}_{mi} r_{mi} + \dot{r}_{mi}^2 \right) \quad (4.47c)$$

for the adaptive multimodal theory.

In the Narimanov-Moiseev asymptotic approximation (4.31a), $F_1^{NM}(t) = F_1(t), F_2^{NM}(t) = F_2(t)$, but the vertical force is computed by

$$F_3^{NM}(t) = -g - \ddot{\eta}_3 - h^{-1} \left(\ddot{p}_{11} p_{11} + \dot{p}_{11}^2 + \ddot{r}_{11} r_{11} + \dot{r}_{11}^2 \right). \quad (4.48)$$

4.4.2 MOMENT

In the third-order adaptive asymptotic approximation, the resulting hydrodynamic moment (relative to the origin O) may be written down ($\boldsymbol{r}_{lC_0} \times \boldsymbol{g}_0 = \boldsymbol{0}$) as

$$\boldsymbol{M}_O = [M_l R_0^2 \sigma^2] \left(\boldsymbol{r}_{lC_0} \times (\boldsymbol{g} - \boldsymbol{g}_0) - \boldsymbol{r}_{lC_0} \times \overset{*}{\boldsymbol{v}}_O + (\boldsymbol{r}_{lC} - \boldsymbol{r}_{lC_0}) \times \boldsymbol{g}_0 \right.$$

$$\left. - \boldsymbol{J}_0^1 \cdot \overset{*}{\boldsymbol{\omega}} - \left(\overset{**}{\boldsymbol{l}}_\omega - \overset{*}{\boldsymbol{l}}_{\omega t} \right) + o(\epsilon) \right)$$

$$= [M_l R_0^2 \sigma^2] \left(F_4(t) \boldsymbol{e}_1(t) + F_5(t) \boldsymbol{e}_2(t) + F_6(t) \boldsymbol{e}_3(t) + o(\epsilon) \right), \quad (4.49)$$

where $\boldsymbol{J}_0^1 = J_{0ij}^1$ is the nondimensional liquid inertia tensor (its $O(1)$-order component),

$$J_{0ij}^1 = \frac{1}{\pi h} \int_{S_0 + \Sigma_0} \Omega_{0i} \frac{\partial \Omega_{0j}}{\partial n} dS, \quad (4.50)$$

within (4.9), $\boldsymbol{g}_0 = -g \boldsymbol{e}_3$ and the following expressions

$$\boldsymbol{l}_\omega = \frac{1}{\pi h} \int_{Q(t)} \boldsymbol{\Omega} \, dQ, \quad \boldsymbol{l}_{\omega t} = \frac{1}{\pi h} \int_{Q(t)} \overset{*}{\boldsymbol{\Omega}} \, dQ \quad (4.51)$$

need to know the generalised Stokes-Joukowski potentials (2.47), (2.48).

The first term, $\boldsymbol{r}_{lC_0} \times (\boldsymbol{g} - \boldsymbol{g}_0)$ in (4.49), implies a quasi-static moment relative to O caused by a small instant pivoting of the tank body by η_4 and η_5. Substituting (2.47), (2.48) into (4.50) computes $\boldsymbol{J}_0^1 = \{J_{0ij}^1\}$ whose the only non-zero component is

$$J_0 = J_{011}^1 = J_{022}^1 = \frac{h^2}{3} - \frac{3}{4} + \frac{8}{h}\sum_n \frac{S_n}{k_{1n}^2(k_{1n}^2 - 1)}, \quad (4.52)$$

where S_n is defined by (4.15). The expression $\overset{**}{\boldsymbol{l}}_\omega - \overset{*}{\boldsymbol{l}}_{\omega t}$ is a rather complicated function of the hydrodynamic generalised coordinates. By utilising the Reynolds transport theorem and the normal relative velocity of the free surface $u_n = \dot{\zeta}/\sqrt{1 + (\nabla \zeta)^2}$, it rewrites in the form

$$\pi h \frac{d^*}{dt}(\overset{*}{\boldsymbol{l}}_\omega - \boldsymbol{l}_{\omega t}) = \frac{d^*}{dt}\int_{\Sigma(t)} \boldsymbol{\Omega}\, u_n\, \mathrm{d}S = \frac{d^*}{dt}\int_0^1 \int_{-\pi}^\pi r\left[\boldsymbol{\Omega}|_{z=\zeta}\right]\frac{\partial \zeta}{\partial t}\mathrm{d}\theta \mathrm{d}r. \quad (4.53)$$

Because $\zeta = O(\epsilon^{1/3})$, getting the $O(\epsilon)$-order component of (4.53) should adopt the $O(\epsilon^{2/3})$-order approximation of the Stokes-Joukowski potentials.

4.4.2.1 Approximate generalised Stokes-Joukowski potentials

The Neumann boundary conditions (2.48) have a specific form for the upright circular base tank when considering them, separately, on the bottom, wetted walls, and the free surface. On the flat bottom, $n_x = n_y = 0, n_z = -1$ and, therefore, the conditions transform to

$$\frac{\partial \Omega_1}{\partial z} = y = r\sin\theta, \quad \frac{\partial \Omega_2}{\partial z} = -x = -r\cos\theta, \quad \frac{\partial \Omega_3}{\partial z} = 0 \text{ at } z = -h; \quad (4.54)$$

on the vertical wall, $n_x = \cos\theta, n_y = \sin\theta, n_z$ and, therefore,

$$\frac{\partial \Omega_1}{\partial r} = -z\sin\theta, \quad \frac{\partial \Omega_2}{\partial r} = z\cos\theta, \quad \frac{\partial \Omega_3}{\partial r} = 0 \text{ at } r = 1. \quad (4.55)$$

The free surface $\Sigma(t)$ is defined by $z = \zeta(r,\theta)$ or, in other words, by the hydrodynamic generalised coordinates in (4.11a). The Neumann boundary conditions on $\Sigma(t)$ take the following form

$$\nabla \Omega_1 \cdot \nabla(z - \zeta) = r\sin\theta + \zeta\left(\sin\theta \frac{\partial \zeta}{\partial r} + \frac{1}{r}\cos\theta \frac{\partial \zeta}{\partial \theta}\right), \quad (4.56a)$$

$$\nabla \Omega_2 \cdot \nabla(z - \zeta) = -r\cos\theta + \zeta\left(-\cos\theta \frac{\partial \zeta}{\partial r} + \frac{1}{r}\sin\theta \frac{\partial \zeta}{\partial \theta}\right), \quad (4.56b)$$

$$\nabla \Omega_3 \cdot \nabla(z - \zeta) = -\frac{\partial \zeta}{\partial \theta}, \quad (4.56c)$$

where

$$\nabla \Omega_i \cdot \nabla(z - \zeta) = \frac{\partial \Omega_i}{\partial z} - \frac{\partial \Omega_i}{\partial r}\frac{\partial \zeta}{\partial r} - \frac{1}{r^2}\frac{\partial \Omega_i}{\partial \theta}\frac{\partial \zeta}{\partial \theta}. \quad (4.57)$$

The forthcoming goal consists, after substituting (4.11a) into the Neumann boundary conditions (4.56), of getting an asymptotic approximation of the Stokes-Joukowski potentials in terms of the hydrodynamic generalised coordinates $p_{Mi}, r_{Mi} = O(\epsilon^{1/3})$,

$$\Omega_{ni} = \Omega_{0i} + \Omega_{1i} + \Omega_{2i} + O(\epsilon), \quad \Omega_{ni} = O(\epsilon^{n/3}); \quad n = 0, 1, 2, \quad (4.58)$$

starting with the zero-order approximation, which suggests the generalised coordinates p_{Mi} and r_{Mi} are zero.

The zero-order approximation Ω_{0i}, $i = 1, 2, 3$ is presented by (4.9). Using the perturbation technique derives the $O(\epsilon^{1/3})$ approximation as follows

$$\Omega_{11} = \sum_{L,k}^{I_\theta, I_r} \mathcal{R}_{Lk}(r) \bar{Z}_{Lk}(z) \cos(L\theta) \sum_{m,i}^{I_\theta, I_r} O^{1,r}_{(Lk),(mi)} r_{mi}(t)$$
$$+ \sum_{l,k}^{I_\theta, I_r} \mathcal{R}_{lk}(r) \bar{Z}_{lk}(z) \sin(l\theta) \sum_{M,i}^{I_\theta, I_r} O^{1,p}_{(lk),(Mi)} p_{Mi}(t), \quad (4.59a)$$

$$\Omega_{12} = \sum_{L,k}^{I_\theta, I_r} \mathcal{R}_{Lk}(r) \bar{Z}_{Lk}(z) \cos(L\theta) \sum_{M,i}^{I_\theta, I_r} O^{2,p}_{(Lk),(Mi)} p_{Mi}(t)$$
$$+ \sum_{l,k}^{I_\theta, I_r} \mathcal{R}_{lk}(r) \bar{Z}_{lk}(z) \sin(l\theta) \sum_{m,i}^{I_\theta, I_r} O^{2,r}_{(lk),(mi)} r_{mi}(t), \quad (4.59b)$$

$$\Omega_{13} = \sum_{m,i}^{I_\theta, I_r} m \mathcal{R}_{mi}(r) \bar{Z}_{mi}(z) \big[\sin(m\theta) p_{mi}(t) - \cos(m\theta) r_{mi}(t) \big], \quad (4.59c)$$

where

$$\bar{Z}_{Mi}(z) = \frac{\cosh(k_{Mi}(z+h))}{k_{Mi} \sinh(k_{Mi} h)} \quad (4.60)$$

and

$$O^{1,r}_{(Lk),(mi)} = \frac{2}{\Lambda_{LL}} \sum_a^\infty \tilde{P}_a \Big\{ \Lambda_{L,m1} \big(\lambda'_{(mi)(1a),(Lk)} - k_{1a}^2 \lambda_{(mi)(1a)(Lk)} \big)$$
$$+ m\Lambda_{mL1,} \bar{\lambda}_{(mi)(1a)(Lk)} \Big\}, \quad (4.61a)$$

$$O^{1,p}_{(lk),(Mi)} = \frac{2}{\Lambda_{ll}} \sum_a^\infty \tilde{P}_a \Big\{ \Lambda_{M,1l} \big(\lambda'_{(Mi)(1a),(lk)} - k_{1a}^2 \lambda_{(Mi)(1a)(lk)} \big)$$
$$- M\Lambda_{1,Ml} \bar{\lambda}_{(Mi)(1a)(lk)} \Big\}; \quad (4.61b)$$

$$O^{2,p}_{(Lk),(Mi)} = \frac{2}{\Lambda_{LL}} \sum_a^\infty \tilde{P}_a \Big[\Lambda_{1ML,} \big(-\lambda'_{(Mi)(1a),(Lk)} + k_{1a}^2 \lambda_{(Mi)(Lk)(1a)} \big)$$
$$- M\Lambda_{L,1M} \bar{\lambda}_{(Mi)(1a)(Lk)} \Big], \quad (4.62a)$$

$$O^{2,r}_{(lk),(mi)} = \frac{2}{\Lambda_{ll}} \sum_a^\infty \tilde{P}_a \Big[\Lambda_{1,lm} \big(-\lambda'_{(mi)(1a),(lk)} + k_{1a}^2 \lambda_{(mi)(lk)(1a)} \big)$$
$$+ m\Lambda_{M,1l} \bar{\lambda}_{(mi)(1a)(lk)} \Big], \quad (4.62b)$$

within

$$\tilde{P}_a = P_a k_{1a}^{-1} \tanh(\tfrac{1}{2} k_{1a} h), \quad (4.63)$$

in which P_a is computed by (4.10); the Λ- and λ-tensors are defined by (4.23), (4.24), and (4.27).

The second-order approximation takes the form

$$\Omega_{21} = \sum_{L,k}^{I_\theta,I_r} \mathcal{R}_{Lk}(r) \bar{\mathcal{Z}}_{Lk}(z) \cos(L\theta) \sum_{Mn,ij}^{I_\theta,I_r} O^{1,pr}_{(Lk),(Mi),(nj)} p_{Mi}(t) r_{nj}(t)$$
$$+ \sum_{l,k}^{I_\theta,I_r} \mathcal{R}_{lk}(r) \bar{\mathcal{Z}}_{lk}(z) \sin(l\theta) \bigg[\sum_{MN,ij}^{I_\theta,I_r} O^{1,pp}_{(lk),(Mi),(Nj)} p_{Mi}(t) p_{Nj}(t)$$
$$+ \sum_{mn,ij}^{I_\theta,I_r} O^{1,rr}_{(lk),(mi),(nj)} r_{mi}(t) r_{nj}(t) \bigg], \quad (4.64a)$$

$$\Omega_{22} = \sum_{L,k}^{I_\theta,I_r} \mathcal{R}_{Lk}(r) \bar{\mathcal{Z}}_{Lk}(z) \cos(L\theta) \bigg[\sum_{MN,ij}^{I_\theta,I_r} O^{2,pp}_{(Lk),(Mi),(Nj)} p_{Mi}(t) p_{Nj}(t)$$
$$+ \sum_{mn,ij}^{I_\theta,I_r} O^{2,rr}_{(Lk),(mi),(nj)} r_{mi}(t) r_{nj}(t) \bigg]$$
$$+ \sum_{l,k}^{I_\theta,I_r} \mathcal{R}_{lk}(r) \bar{\mathcal{Z}}_{lk}(z) \sin(l\theta) \sum_{Mn,ij}^{I_\theta,I_r} O^{2,pr}_{(lk),(Mi),(nj)} p_{Mi}(t) r_{nj}(t), \quad (4.64b)$$

$$\Omega_{23} = \sum_{L,k}^{I_\theta,I_r} \mathcal{R}_{Lk}(r) \bar{\mathcal{Z}}_{Lk}(z) \cos(L\theta) \bigg[\sum_{Mn,ij}^{I_\theta,I_r} O^{3,pr}_{(Lk),(Mi),(nj)} p_{Mi}(t) r_{nj}(t) \bigg]$$
$$+ \sum_{l,k}^{I_\theta,I_r} \mathcal{R}_{lk}(r) \bar{\mathcal{Z}}_{lk}(z) \sin(l\theta) \bigg[\sum_{MN,ij}^{I_\theta,I_r} O^{3,pp}_{(lk),(Mi),(Nj)} p_{Mi}(t) p_{Nj}(t)$$

$$+ \sum_{mn,ij}^{I_\theta,I_r} O^{3,rr}_{(lk),(mi),(nj)} r_{mi}(t) r_{nj}(t) \Bigg], \quad (4.64c)$$

where

$$O^{1,pr}_{(Lk),(Mi),(nj)} = \frac{1}{\Lambda_{LL}} \Bigg\{ 2 \sum_a^\infty P_a \Big[\Lambda_{LM,1n}(\lambda'_{(nj)(1a),(Mi)(Lk)} + \lambda'_{(Mi)(1a),(nj)(Lk)}$$
$$- k_{1a}^2 \lambda_{(nj)(1a)(Mi)(Lk)}) + (n\Lambda_{LMn1} - M\Lambda_{L1,Mn})\bar{\lambda}_{(nj)(1a)(Mi)(Lk)} \Big]$$
$$+ \sum_{A,b}^{I_\theta,I_r} \kappa_{Ab}^{-1} \Big[O^{1,p}_{(Ab),(Mi)}(\Lambda_{L,An}[\lambda'_{(Ab)(nj),(Lk)} - k_{Ab}^2 \lambda_{(Ab)(nj)(Lk)}]$$
$$+ An\Lambda_{AnL,}\bar{\lambda}_{(Ab)(nj)(Lk)}) + O^{1,r}_{(Ab),(nj)}(\Lambda_{AML,}[\lambda'_{(Ab)(Mi),(Lk)}$$
$$- k_{Ab}^2 \lambda_{(Ab)(Mi)(Lk)}] + AM\Lambda_{L,AM}\bar{\lambda}_{(Ab)(Mi)(Lk)}) \Big] \Bigg\}, \quad (4.65a)$$

$$O^{1,pp}_{(lk),(Mi),(Nj)} = \frac{1}{\Lambda_{ll}} \Bigg\{ 2 \sum_a^\infty P_a \Big[\Lambda_{MN,l1}(\lambda'_{(Nj(1a),(Mi)(lk)} - \tfrac{1}{2} k_{1a}^2 \lambda_{(Mi)(1a)(Nj)(lk)})$$
$$- N\Lambda_{1M,lN}\bar{\lambda}_{(mi)(1a)(Nj)(lk)} \Big] + \sum_{A,b}^{I_\theta,I_r} O^{1,p}_{(Ab),(Mi)} \kappa_{Ab}^{-1} \Big[\Lambda_{N,Al}(\lambda'_{(Ab)(Nj),(lk)}$$
$$- k_{Ab}^2 \lambda_{(Ab)(Nj)(lk)}) - AN\Lambda_{A,Nl}\bar{\lambda}_{(Ab)(Nj)(lk)} \Big] \Bigg\}, \quad (4.65b)$$

$$O^{1,rr}_{(lk),(mi),(nj)} = \frac{1}{\Lambda_{ll}} \Bigg\{ 2 \sum_a^\infty P_a \Big[\Lambda_{,1nml}(\lambda'_{(nj)(1a),(mi)(lk)} - \tfrac{1}{2} k_{1a}^2 \lambda_{(mi)(1a)(nj)(lk)})$$
$$+ n\Lambda_{1n,ml}\bar{\lambda}_{(mi)(1a)(nj)(lk)} \Big] + \sum_{A,b}^{I_\theta,I_r} O^{1,r}_{(Ab),(mi)} \kappa_{Ab}^{-1} \Big[\Lambda_{A,nl}(\lambda'_{(Ab)(nj),(lk)}$$
$$- k_{Ab}^2 \lambda_{(Ab)(nj)(lk)}) - An\Lambda_{n,Al}\bar{\lambda}_{(Ab)(nj)(lk)} \Big] \Bigg\}; \quad (4.65c)$$

$$O^{2,pp}_{(Lk),(Mi),(Nj)} = \frac{1}{\Lambda_{LL}} \Bigg\{ 2 \sum_a^\infty P_a \Big[\Lambda_{1MNL,}(\tfrac{1}{2} k_{1a}^2 \lambda_{(Mi)(Nj)(1a)(Lk)}$$
$$- \lambda'_{(Nj)(1a),(Mi)(Lk)}) - N\Lambda_{ML,N1}\bar{\lambda}_{(Nj)(1a)(Mi)(Lk)} \Big] + \sum_{A,b}^{I_\theta,I_r} O^{2,p}_{(Ab),(Mi)} \kappa_{Ab}^{-1}$$

$$\times \left[\Lambda_{ANL,}(\lambda'_{(Ab)(Nj),(Lk)} - k^2_{Ab}\lambda_{(Ab)(Nj)(Lk)}) + AN\Lambda_{L,AN}\bar{\lambda}_{(Ab)(Nj)(Lk)} \right] \Big\}, \tag{4.66a}$$

$$O^{2,rr}_{(Lk),(mi),(nj)} = \frac{1}{\Lambda_{LL}}\Bigg\{ 2\sum_{a}^{\infty} P_a \Big[\Lambda_{1L,mn}\big(\tfrac{1}{2}k^2_{1a}\lambda_{(mi)(nj)(1a)(Lk)}$$
$$-\lambda'_{(nj)(1a),(mi)(Lk)}\big) + n\Lambda_{nl,1m}\bar{\lambda}_{(nj)(1a)(mi)(Lk)}\Big] + \sum_{a,b}^{I_\theta,I_r} O^{2,r}_{(ab),(mi)}\kappa^{-1}_{ab}$$
$$\times\Big[\Lambda_{L,an}\big(\lambda'_{(ab)(nj),(Lk)} - k^2_{ab}\lambda_{(ab)(nj)(Lk)}\big) + an\,\Lambda_{anL,}\bar{\lambda}_{(ab)(nj)(Lk)}\Big]\Bigg\}, \tag{4.66b}$$

$$O^{2,pr}_{(lk),(Mi),(nj)} = \frac{1}{\Lambda_{ll}}\Bigg\{2\sum_{a}^{\infty} P_a\Big[\Lambda_{1M,ln}\big(k^2_{1a}\lambda_{(Mi)(1a)(nj)(lk)} - \lambda'_{(nj)(1a),(Mi)(lk)}$$
$$-\lambda'_{(Mi)(1a),(nj)(lk)}\big) + (n\Lambda_{Mn,1l} - M\Lambda_{,1Mnl})\bar{\lambda}_{(1a)(Mi)(nj)(lk)}\Big]$$
$$+\sum_{Ab}^{I_\theta,I_r}\kappa^{-1}_{Ab}\Big[O^{2,p}_{(Ab),(Mi)}\big(\Lambda_{A,nl}(\lambda'_{(Ab)(nj),(lk)} - k^2_{Ab}\lambda_{(Ab)(nj)(lk)})$$
$$- An\Lambda_{n,Al}\bar{\lambda}_{(Ab)(nj)(lk)}\big) + O^{2,r}_{(Ab),(nj)}\big(\Lambda_{M,lA}(\lambda'_{(Ab)(Mi),(lk)} - k^2_{Ab}\lambda_{(Ab)(Mi)(lk)})$$
$$- AM\Lambda_{A,Ml}\bar{\lambda}_{(Ab)(Mi)(lk)}\big)\Big]\Bigg\}; \tag{4.66c}$$

$$O^{3,pr}_{(Lk),(Mi),(nj)} = \frac{1}{\Lambda_{LL}}\Bigg[\lambda_{(Mi)(nj)(Lk)}\bigg(-\Lambda_{L,Mn}\frac{Mk^2_{Mi}}{\kappa_{Mi}} + \Lambda_{MLn,}\frac{nk^2_{nj}}{\kappa_{nj}}\bigg)$$
$$+\lambda'_{(Mi)(nj),(Lk)}\big(\Lambda_{L,Mn}M\kappa^{-1}_{Mi} - \Lambda_{LMn,}n\kappa^{-1}_{nj}\big)$$
$$+\bar{\lambda}_{(Mi)(nj)(Lk)}Mn\big(\Lambda_{LMn,}M\kappa^{-1}_{Mi} - \Lambda_{L,Mn}n\kappa^{-1}_{nj}\big)\Bigg], \tag{4.67a}$$

$$O^{3,pp}_{(lk),(Mi),(Nj)} = \frac{1}{\Lambda_{ll}}\frac{M}{\kappa_{Mi}}\Bigg[\Lambda_{N,Ml}\big(-k^2_{Mi}\lambda_{(Mi)(Nj)(lk)} + \lambda'_{(Mi)(Nj),(lk)}\big)$$
$$- MN\,\Lambda_{M,Nl}\bar{\lambda}_{(Mi)(Nj)(lk)}\Bigg], \tag{4.67b}$$

$$O^{3,rr}_{(lk),(mi),(nj)} = \frac{1}{\Lambda_{ll}}\frac{m}{\kappa_{mi}}\Bigg[\Lambda_{m,nl}\big(k^2_{mi}\lambda_{(mi)(nj)(lk)} - \lambda'_{(mi)(nj),(lk)}\big)$$
$$+ mn\,\Lambda_{n,ml}\bar{\lambda}_{(mi)(nj)(lk)}\Bigg]. \tag{4.67c}$$

4.4.2.2 Adaptive approximation of the hydrodynamic moment

Based on the Stokes-Joukowski potentials from the previous section, the moment component $F_4(t)$ is equal to

$$F_4(t) = g\left[\tfrac{1}{2}h\,\eta_4 - h^{-1}\sum_a^{I_r} P_a r_{1a}\right] - \tfrac{1}{2}h\ddot{\eta}_2 - J_0\ddot{\eta}_4 - 2h^{-1}\sum_a^{I_r} \tilde{P}_a \ddot{r}_{1a}$$

$$- \frac{1}{\pi h}\frac{d}{dt}\left\{\sum_{Nm,ij}^{I_\theta,I_r} \tilde{O}^{1,pr}_{(Nj),(mi)}\dot{p}_{Nj}r_{mi} + \sum_{nM,ij}^{I_\theta,I_r} \tilde{O}^{1,rp}_{(nj),(Mi)}\dot{r}_{nj}p_{Mi}\right.$$

$$+ \sum_{NmL,ijk}^{I_\theta,I_r} \tilde{O}^{1,ppr}_{(Nj),(Lk),(mi)}\dot{p}_{Nj}p_{Lk}r_{mi} + \sum_{nML,ijk}^{I_\theta,I_r} \tilde{O}^{1,rpp}_{(nj),(Lk),(Mi)}\dot{r}_{nj}p_{Lk}p_{Mi}$$

$$\left. + \sum_{nml,ijk}^{I_\theta,I_r} \tilde{O}^{1,rrr}_{(nj),(lk),(mi)}\dot{r}_{nj}r_{lk}r_{mi}\right\}, \quad (4.68)$$

where

$$\tilde{O}^{1,pr}_{(Nj),(mi)} = \Lambda_{N,1m}\sum_a^{\infty} P_a\lambda_{(1a)(mi)(Nj)} + \Lambda_{NN}\kappa_{Nj}^{-1}O^{1,r}_{(Nj),(mi)},$$

$$\tilde{O}^{1,rp}_{(nj),(Mi)} = \Lambda_{M,1n}\sum_a^{\infty} P_a\lambda_{(1a)(Mi)(nj)} + \Lambda_{nn}\kappa_{nj}^{-1}O^{1,p}_{(nj),(Mi)},$$

$$\tilde{O}^{1,ppr}_{(Nj),(Lk),(mi)} = 2\Lambda_{NL,m1}\sum_a^{\infty} \tilde{P}_a k_{1a}^2 \lambda_{(1a)(Nj)(Lk)(mi)}$$

$$+ \Lambda_{NN}\kappa_{Nj}^{-1}O^{1,pr}_{(Nj),(Lk),(mi)} + \sum_{A,b}^{I_\theta,I_r}\lambda_{(Ab)(Lk)(Nj)}O^{1,r}_{(Ab),(mi)}\Lambda_{ALN},$$

$$+ \sum_{a,b}^{I_\theta,I_r}\lambda_{(ab)(mi)(Nj)}O^{1,p}_{(ab),(Lk)}\Lambda_{N,ma},$$

$$\tilde{O}^{1,rpp}_{(nj),(Lk),(Mi)} = \Lambda_{LM,n1}\sum_a^{\infty} \tilde{P}_a k_{1a}^2 \lambda_{(1a)(nj)(Mi)(Lk)}$$

$$+ \Lambda_{nn}\kappa_{nj}^{-1}O^{1,pp}_{(nj),(Mi),(Lk)} + \sum_{a,b}^{I_\theta,I_r}\lambda_{(ab)(Lk)(nj)}O^{1,p}_{(ab),(Mi)}\Lambda_{L,an},$$

$$\tilde{O}^{1,rrr}_{(nj),(lk),(mi)} = \Lambda_{,1nlm}\sum_a^{\infty} \tilde{P}_a k_{1a}^2 \lambda_{(1a)(nj)(mi)(lk)}$$

$$+ \Lambda_{nn}\kappa_{nj}^{-1}O^{1,rr}_{(nj),(mi),(lk)} + \sum_{A,b}^{I_\theta,I_r}\lambda_{(Ab)(lk)(nj)}O^{1,r}_{(Ab),(mi)}\Lambda_{A,ln}.$$

The moment component $F_5(t)$ reads as

$$F_5(t) = g\left[\tfrac{1}{2}h\eta_5 + h^{-1}\sum_a^{I_r} P_a p_{1a}\right] + \tfrac{1}{2}h\ddot{\eta}_1 - J_0\ddot{\eta}_5 + 2h^{-1}\sum_a^{I_r} \tilde{P}_a \ddot{p}_{1a}$$

$$-\frac{1}{\pi h}\frac{d}{dt}\Bigg\{\sum_{NM,ij}^{I_\theta,I_r} \tilde{O}^{2,pp}_{(Nj),(Mi)}\dot{p}_{Nj}p_{mi} + \sum_{mn,ij}^{I_\theta,I_r} \tilde{O}^{2,rr}_{(nj),(mi)}\dot{r}_{nj}r_{mi}$$

$$+ \sum_{NML,ijk}^{I_\theta,I_r} \tilde{O}^{2,ppp}_{(Nj),(Lk),(Mi)}\dot{p}_{Nj}p_{Lk}p_{Mi} + \sum_{nml,ijk}^{I_\theta,I_r} \tilde{O}^{2,prr}_{(Nj),(lk),(mi)}\dot{p}_{Nj}r_{lk}r_{mi}$$

$$+ \sum_{nmL,ijk}^{I_\theta,I_r} \tilde{O}^{2,rpr}_{(nj),(Lk),(mi)}\dot{r}_{nj}p_{Lk}r_{mi}\Bigg\}, \quad (4.69)$$

where

$$\tilde{O}^{2,pp}_{(Nj),(Mi)} = -\Lambda_{1MN,}\sum_a^\infty P_a\lambda_{(1a)(Mi)(Nj)} + \Lambda_{NN}\kappa_{Nj}^{-1}O^{2,p}_{(Nj),(Mi)},$$

$$\tilde{O}^{2,rr}_{(nj),(mi)} = -\Lambda_{1,mn}\sum_a^\infty P_a\lambda_{(1a)(mi)(nj)} + \Lambda_{nn}\kappa_{nj}^{-1}O^{2,r}_{(nj),(mi)},$$

$$\tilde{O}^{2,ppp}_{(Nj),(Lk),(Mi)} = -\Lambda_{NLM1,}\sum_a^\infty \tilde{P}_a k_{1a}^2\lambda_{(1a)(Nj)(Lk)(Mi)}$$

$$+ \Lambda_{NN}\kappa_{Nj}^{-1}O^{2,pp}_{(Nj),(Mi),(Lk)} + \sum_{A,b}^{I_\theta,I_r}\lambda_{(Ab)(Lk)(Nj)}O^{2,p}_{(Ab),(Mi)}\Lambda_{ALN,,}$$

$$\tilde{O}^{2,prr}_{(Nj),(lk),(mi)} = -\Lambda_{N1,lm}\sum_a^\infty \tilde{P}_a k_{1a}^2\lambda_{(1a)(Nj)(mi)(lk)}$$

$$+ \Lambda_{NN}\kappa_{Nj}^{-1}O^{2,rr}_{(Nj),(mi),(lk)} + \sum_{a,b}^{I_\theta,I_r}\lambda_{(ab)(lk)(Nj)}O^{2,r}_{(ab),(mi)}\Lambda_{N,al},$$

$$\tilde{O}^{2,rpr}_{(nj),(Lk),(mi)} = -2\Lambda_{L1,nm}\sum_a^\infty \tilde{P}_a k_{1a}^2\lambda_{(1a)(Lk)(mi)(nj)}$$

$$+ \Lambda_{nn}\kappa_{nj}^{-1}O^{2,pr}_{(nj),(Lk),(mi)} + \sum_{A,b}^{I_\theta,I_r}\Lambda_{A,mn}O^{2,p}_{(Ab),(Lk)}\lambda_{(Ab)(mi)(nj)}$$

$$+ \sum_{a,b}^{I_\theta,I_r}\Lambda_{L,an}O^{2,r}_{(ab),(mi)}\lambda_{(ab)(Lk)(nj)}.$$

The hydrodynamic moment component $F_6(t)$ is only function of (4.51), i.e.,

$$F_6(t) = -\frac{1}{\pi h}\frac{d}{dt}\int_0^1\int_{-\pi}^{\pi} r\,\Omega_3|_{z=\zeta}\,\partial_t\zeta\,d\theta dr$$

$$= -\frac{1}{h}\frac{d}{dt}\sum_{m,i}^{I_\theta,I_r} m\kappa_{mi}^{-1}(p_{mi}\dot{r}_{mi}-\dot{p}_{mi}r_{mi})-\frac{1}{\pi h}\frac{d}{dt}\Bigg\{\sum_{nML,ijk}^{I_\theta,I_r}\tilde{O}^{3,ppr}_{(Lk),(Mi),(nj)}\dot{p}_{Lk}p_{Mi}r_{nj}$$

$$+\sum_{nMl,ijk}^{I_\theta,I_r}\tilde{O}^{3,rpp}_{(lk),(Mi),(nj)}\dot{r}_{lk}p_{Mi}p_{Nj}+\sum_{nml,ijk}^{I_\theta,I_r}\tilde{O}^{3,rrr}_{(lk),(mi),(nj)}\dot{r}_{lk}r_{mi}r_{nj}\Bigg\}, \quad (4.70)$$

where

$$\tilde{O}^{3,ppr}_{(Lk),(Mi),(nj)} = \frac{\Lambda_{LL}}{\kappa_{Lk}}O^{3,pr}_{(Lk),(Mi),(nj)} + \lambda_{(Lk)(Mi)(nj)}(M\Lambda_{L,Mn}-n\Lambda_{LMn,}),$$

$$\tilde{O}^{3,rpp}_{(lk),(Mi),(Nj)} = \Lambda_{ll}\kappa_{lk}^{-1}O^{3,pp}_{(lk),(Mi),(Nj)} + \lambda_{(lk)(Mi)(Nj)}M\Lambda_{N,lM},$$

$$\tilde{O}^{3,rrr}_{(lk),(mi),(nj)} = \Lambda_{ll}\kappa_{lk}^{-1}O^{3,rr}_{(lk),(mi),(nj)} - \lambda_{(lk)(mi)(nj)}m\Lambda_{m,ln}.$$

4.4.2.3 Narimanov-Moiseev approximation

Adopting the Narimanov-Moiseev asymptotic relations (4.31a) and neglecting the $o(\epsilon)$-order quantities in formulas for the hydrodynamic moment from the previous section derives

$$F_4(t) = g\Big[\tfrac{1}{2}h\,\eta_4 - h^{-1}\sum_a^{I_r} P_a r_{1a}\Big] - \tfrac{1}{2}h\ddot{\eta}_2 - J_0\ddot{\eta}_4 - 2h^{-1}\sum_a^{I_r}\tilde{P}_a\ddot{r}_{1a}$$

$$-\frac{1}{\pi h}\frac{d}{dt}\Bigg\{\mu_1\dot{p}_{11}p_{11}r_{11} + \dot{r}_{11}\left(-\mu_2 p_{11}^2 + \mu_3 r_{11}^2\right) + \sum_i^{I_r}\Big[\mu_4^{(i)}(p_{11}\dot{r}_{2i}-\dot{r}_{11}p_{2i})$$

$$+\mu_5^{(i)}(p_{11}\dot{r}_{2i}-r_{11}\dot{p}_{2i}) + \mu_6^{(i)}r_{11}\dot{p}_{0i} + \mu_7^{(i)}\dot{r}_{11}p_{0i}\Big]\Bigg\}, \quad (4.71a)$$

$$F_5(t) = g\Big[\tfrac{1}{2}h\eta_5 + h^{-1}\sum_a^{I_r} P_a p_{1a}\Big] + \tfrac{1}{2}h\ddot{\eta}_1 - J_0\ddot{\eta}_5 + 2h^{-1}\sum_a^{I_r}\tilde{P}_a\ddot{p}_{1a}$$

$$-\frac{1}{\pi h}\frac{d}{dt}\Bigg\{-\mu_1\dot{r}_{11}p_{11}r_{11} + \dot{p}_{11}\left(\mu_2 r_{11}^2-\mu_3 p_{11}^2\right)-\sum_i^{I_r}\Big[\mu_4^{(i)}(\dot{p}_{11}p_{2i}+\dot{r}_{11}r_{2i})$$

$$+\mu_5^{(i)}(p_{11}\dot{p}_{2i}+r_{11}\dot{r}_{2i}) + \mu_6^{(i)}p_{11}\dot{p}_{0i} + \mu_7^{(i)}p_{0i}\dot{p}_{11}\Big]\Bigg\}, \quad (4.71b)$$

$$F_6(t) = -\frac{1}{h\kappa_{11}}\frac{d}{dt}(\dot{r}_{11}p_{11}-\dot{p}_{11}r_{11}), \quad (4.71c)$$

where

$$\mu_1 = \tilde{O}^{1,ppr}_{(11),(11),(11)} = -\tilde{O}^{2,rpr}_{(11),(11),(11)},\ \mu_2 = \tilde{O}^{2,prr}_{(11),(11),(11)} = -\tilde{O}^{1,rpp}_{(11),(11),(11)},$$

$$\mu_3 = \tilde{O}^{1,rrr}_{(11),(11),(11)} = -\tilde{O}^{2,ppp}_{(11),(11),(11)},$$

$$\mu_4^{(i)} = \tilde{O}^{1,pr}_{(11),(2i)} = -\tilde{O}^{1,rp}_{(11),(2i)} = -\tilde{O}^{2,pp}_{(11),(2i)} = -\tilde{O}^{2,rr}_{(11),(2i)},$$

$$\mu_5^{(i)} = \tilde{O}^{1,rp}_{(2i),(11)} = -\tilde{O}^{1,pr}_{(2i),(11)} = -\tilde{O}^{2,pp}_{(2i),(11)} = -\tilde{O}^{2,rr}_{(2i),(11)},$$

$$\mu_6^{(i)} = \tilde{O}^{1,pr}_{(0i),(11)} = -\tilde{O}^{2,pp}_{(0i),(11)}, \quad \mu_7^{(i)} = \tilde{O}^{1,rp}_{(11),(0i)} = -\tilde{O}^{2,pp}_{(11),(0i)}.$$

The hydrodynamic coefficients at the nonlinear terms in formulas (4.71) are functions of the nondimensional liquid depth h. Tables 8.18 and 8.19 list these coefficients versus for $h \geq 1.05$.

Remark 4.5. *The yaw moment component (4.71c) is zero in the linear approximation. This looks rather expectable for the chosen position of the coordinate system and the axisymmetric mean liquid domain. However, this component can formally be non-zero in the nonlinear approximation due to the quantity $d(\dot{r}_{11}p_{11} - \dot{p}_{11}r_{11})/dt$.*

5 Narimanov-Moiseev–type equations and resonant steady-state waves

5.1 ASYMPTOTICALLY EQUIVALENT ORBITAL TANK MOTIONS

We consider a small-magnitude periodic motion of a circular base tank with four degrees of freedom associated with surge, roll, sway, and pitch, which, in the nondimensional analysis, are governed by the 2π-periodic functions

$$\eta_i(t) = \sum_{k=1}^{\infty} \left[\eta_{ia}^{(k)} \cos kt + \mu_{ia}^{(k)} \sin kt \right], \quad \eta_{ia}^{(k)} \sim \mu_{ia}^{(k)} = O(\epsilon), \; i = 1, 2, 4, 5. \quad (5.1)$$

Because of the resonance condition for the lowest natural sloshing mode, coefficients at the lowest Fourier harmonics in (5.1) should not be zero, simultaneously, i.e.,

$$O(\epsilon) = \sum_{i=1,2,4,5} |\eta_{ia}^{(1)}| + |\mu_{ia}^{(1)}| \neq 0. \quad (5.2)$$

In order to describe all admissible steady-state resonant waves, one should construct a general asymptotic periodic solution of the Narimanov-Moiseev–type modal system (4.32)–(4.35), which is based on the asymptotic relations (4.31) postulates that $p_{11}(t), r_{11}(t) = O(\epsilon^{1/3})$ and $p_{0k}(t), p_{2k}(t), r_{2k}(t) = O(\epsilon^{2/3}), k = 1, 2, \ldots$. Specifically, the third-order hydrodynamic generalised coordinates $p_{3k}(t), r_{3k}(t), p_{1n}(t), r_{1n}(t), k \geqslant 1$, and $n \geqslant 2$ are absent in (4.32)–(4.33); they are 'driven' by $p_{11}(t), r_{11}(t)$, and $p_{0k}(t), p_{2k}(t), r_{2k}(t)$ and can easily be found from equations (4.34), (4.35) if we know the lower-order hydrodynamic generalised coordinates, $O(\epsilon^{1/3})$ and $O(\epsilon^{2/3})$. Concentrating on the $O(\epsilon^{1/3})$ and $O(\epsilon^{2/3})$ asymptotic components of the periodic asymptotic solution suggests dealing with the system (4.32), (4.33), which has inhomogeneous terms only in the right-sides of (4.32). Substitution of (5.1) into these right-hand sides gives

$$P_1 \kappa_{11} \sum_{k=1}^{\infty} \left[(k\eta_{1a}^{(k)} - (kS_1 - g)\eta_{5a}^{(k)}) \cos kt + (k\mu_{1a}^{(k)} - (kS_1 - g)\mu_{5a}^{(k)}) \sin kt \right],$$

$$P_1 \kappa_{11} \sum_{k=1}^{\infty} \left[(k\eta_{2a}^{(k)} + (kS_1 - g)\eta_{4a}^{(k)}) \cos kt + (k\mu_{2a}^{(k)} + (kS_1 - g)\mu_{4a}^{(k)}) \sin kt \right], \quad (5.3)$$

where, by using the Moiseev detuning (4.31b) and neglecting the $o(\epsilon)$-order quantities, one can replace the nondimensional gravity acceleration g in (4.1) by $g := g/(R_0 \sigma_{11}^2)$.

The Narimanov-Moiseev asymptotic theory implies that only the first Fourier harmonics in $\eta_i(t)$, being of the $O(\epsilon)$-order, resonantly perturbs the lowest-order generalised coordinates p_{11} and r_{11} up to the order $O(\epsilon^{1/3})$, while other (higher) Fourier harmonics in the right-hand side of (4.32) (= (5.3)) contribute the $O(\epsilon)$-order component to r_{11} and p_{11}. Concentrating on this lowest Fourier harmonics, $\cos t$ and $\sin t$, in (5.3), we get (5.3) rewritten in the form $\epsilon_x \cos t + \bar{\epsilon}_x \sin t$ and $\bar{\epsilon}_y \cos t + \epsilon_y \sin t$, where

$$\epsilon_x = P_1 \kappa_{11} \left(\eta_{1a}^{(1)} - [S_1 - g]\eta_{5a}^{(1)} \right), \quad \bar{\epsilon}_x = P_1 \kappa_{11} \left(\mu_{1a}^{(1)} - [S_1 - g]\mu_{5a}^{(1)} \right),$$
$$\bar{\epsilon}_y = P_1 \kappa_{11} \left(\eta_{2a}^{(1)} + [S_1 - g]\eta_{4a}^{(1)} \right), \quad \epsilon_y = P_1 \kappa_{11} \left(\mu_{2a}^{(1)} + [S_1 - g]\mu_{4a}^{(1)} \right). \tag{5.4}$$

The same single-harmonic right-hand sides in (4.32) appear for the orbital horizontal translational tank motions

$$\eta_1(t) = e_x \cos t + \bar{e}_x \sin t = \frac{1}{P_1 \kappa_{11}} (\epsilon_x \cos t + \bar{\epsilon}_x \sin t),$$
$$\eta_2(t) = \bar{e}_y \cos t + e_y \sin t = \frac{1}{P_1 \kappa_{11}} (\bar{\epsilon}_y \cos t + \epsilon_y \sin t). \tag{5.5}$$

When $e_x e_y - \bar{e}_x \bar{e}_y = 0$, (5.5) implies a harmonic horizontal tank reciprocation. If $e_x e_y - \bar{e}_x \bar{e}_y \neq 0$, (5.5) determines a centred elliptic curve in the Oxy-plane, which is governed by

$$(\bar{e}_y^2 + e_y^2)x^2 + (e_x^2 + \bar{e}_x^2)x^2 - 2(e_x \bar{e}_y + \bar{e}_x e_y)xy = (e_x e_y - \bar{e}_x \bar{e}_y)^2 \tag{5.6}$$

(readers can prove that by computing the fundamental invariants of (5.6)). The elliptic orbit direction depends on the sign of $e_x e_y - \bar{e}_x \bar{e}_y$. The positive sign implies counterclockwise orbit but the negative sign – clockwise.

In summary, we showed that *any periodic asymptotic solution of the Narimanov-Moiseev-type modal system for the combined resonant periodic sway, roll, pitch, and surge has an asymptotically equivalent steady-state solution, which is associated with a resonant horizontal translational orbital tank motion whose particular cases are the reciprocation and the counterclockwise/clockwise elliptic orbital tank movement.* The asymptotic equivalence means that both asymptotic solutions have identical first- and second-order wave components, and the solutions possess the same stability properties. The latter fact will be proven in section 5.2.2.

Remark 5.1. *Rotating Oxy around the Oz axis to superpose Ox with the major axis of the centred ellipse transforms (5.6) to the canonic form $x^2/e_x'^2 + y^2/e_y'^2 = 1$, where $|e_x'|$ and $|e_y'|$ are the semi-axes of the elliptic trajectory computed from the quadratic form (5.6). For this position of the Oxy frame, (5.5) can be rewritten in the form*

$$\eta_1(t) = e_x' \cos t, \quad \eta_2(t) = e_y' \sin t, \tag{5.7}$$

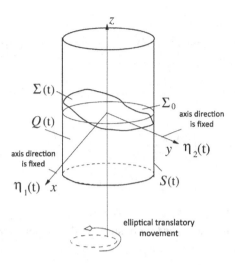

Figure 5.1 The tank performs an orbital (elliptical) translational motion in the counterclockwise direction; Oz is the axis of symmetry. The liquid domain $Q(t)$ is bounded by the free surface $\Sigma(t)$ and the wetted tank surface $S(t)$.

where, without loss of generality, $e'_x > 0$, $|e'_y| \leq e'_x$, but e'_y is positive for counterclockwise elliptic orbits and negative for clockwise ones. Proceeding with (5.7) and $\eta_4 = \eta_5 = 0$ in (4.32) derives the right-hand sides $\epsilon'_x \cos t$ and $\epsilon'_y \sin t$ in (4.32a) and (4.32b), respectively, where

$$\epsilon'_x = P_1 \kappa_{11} e'_x > 0, \quad \epsilon'_y = P_1 \kappa_{11} e'_y, \quad 0 \leq |\epsilon'_y| \leq \epsilon'_x. \tag{5.8}$$

Because of the axial symmetry, the forthcoming analysis may, without loss of generality, concentrate on counterclockwise orbits with $\epsilon'_y \geq 0$. This elliptic orbital forcing is illustrated in Figure 5.1.

5.2 PERIODIC SOLUTIONS OF THE NARIMANOV-MOISEEV–TYPE MODAL EQUATIONS

5.2.1 THE LOWEST-ORDER TERMS; SECULAR EQUATIONS

In order to find the general asymptotic periodic solution of the Narimanov-Moiseev-type equations, we use the Bubnov-Galerkin method [52]. For this purpose, we write down the first-order asymptotic contribution in the form

$$p_{11}(t) = a \cos t + \bar{a} \sin t + O(\epsilon), \quad r_{11}(t) = \bar{b} \cos t + b \sin t + O(\epsilon), \tag{5.9}$$

where the nondimensional amplitude parameters $a, \bar{a}, \bar{b}$, and $b = O(\epsilon^{1/3})$. Substituting (5.9) into (4.33) derives

$$p_{0k}(t) = s_{0k}(a^2 + \bar{a}^2 + b^2 + \bar{b}^2)$$
$$+ s_{1k}\left[(a^2 - \bar{a}^2 - b^2 + \bar{b}^2)\cos 2t + 2(a\bar{a} + b\bar{b})\sin 2t\right] + o(\epsilon), \tag{5.10a}$$

$$p_{2k}(t) = c_{0k}(a^2 + \bar{a}^2 - b^2 - \bar{b}^2)$$
$$+ c_{1k}\left[(a^2 - \bar{a}^2 + b^2 - \bar{b}^2)\cos 2t + 2(a\bar{a} - b\bar{b})\sin 2t\right] + o(\epsilon), \quad (5.10b)$$

$$r_{2k}(t) = 2c_{0k}(a\bar{b} + b\bar{a}) + 2c_{1k}\left[(a\bar{b} - b\bar{a})\cos 2t + (ab + \bar{a}\bar{b})\sin 2t\right] + o(\epsilon), \quad (5.10c)$$

where

$$s_{0k} = \tfrac{1}{2}\left(\frac{d_{10,k} - d_{8,k}}{\bar{\sigma}_{0k}^2}\right), \quad s_{1k} = \frac{d_{10,k} + d_{8,k}}{2(\bar{\sigma}_{0k}^2 - 4)},$$
$$c_{0k} = \tfrac{1}{2}\left(\frac{d_{9,k} - d_{7,k}}{\bar{\sigma}_{2k}^2}\right), \quad c_{1k} = \frac{d_{9,k} + d_{7,k}}{2(\bar{\sigma}_{2k}^2 - 4)}. \quad (5.11)$$

Substituting (5.9) and (5.10) into (4.32) and grouping the terms in front of the first harmonic, for $\cos t$ and $\sin t$, deduces the *necessary* solvability (secular) condition

$$\begin{cases} \text{①}: a\left[\Lambda + m_1(a^2 + \bar{a}^2 + \bar{b}^2) + m_3 b^2\right] + \bar{a}[(m_1 - m_3)\bar{b}b + \xi] = \epsilon_x, \\ \text{②}: \bar{a}\left[\Lambda + m_1(a^2 + \bar{a}^2 + b^2) + m_3 \bar{b}^2\right] + a[(m_1 - m_3)\bar{b}b - \xi] = \bar{\epsilon}_x, \\ \text{③}: b\left[\Lambda + m_1(b^2 + \bar{b}^2 + \bar{a}^2) + m_3 a^2\right] + \bar{b}[(m_1 - m_3)\bar{a}a - \xi] = \epsilon_y, \\ \text{④}: \bar{b}\left[\Lambda + m_1(b^2 + \bar{b}^2 + a^2) + m_3 \bar{a}^2\right] + b[(m_1 - m_3)\bar{a}a + \xi] = \bar{\epsilon}_y, \end{cases} \quad (5.12)$$

with respect to $a, \bar{a}, \bar{b}$, and b. Here, $\xi = 2\xi_{11}$; the right-hand side is computed by (5.4), and the coefficients m_1 and m_3 are calculated by the formulas

$$m_1 = -\tfrac{1}{2}d_1 + \sum_{j=1}^{I_r}\left[c_{1j}\left(\tfrac{1}{2}d_3^{(j)} - 2d_4^{(j)}\right) + s_{1j}\left(\tfrac{1}{2}d_5^{(j)} - 2d_6^{(j)}\right) - s_{0j}d_5^{(j)} - c_{0j}d_3^{(j)}\right],$$
(5.13a)

$$m_3 = \tfrac{1}{2}d_1 - 2d_2$$
$$+ \sum_{j=1}^{I_r}\left[c_{1j}\left(\tfrac{3}{2}d_3^{(j)} - 6d_4^{(j)}\right) + s_{1j}\left(-\tfrac{1}{2}d_5^{(j)} + 2d_6^{(j)}\right) - s_{0j}d_5^{(j)} + c_{0j}d_3^{(j)}\right], \quad (5.13b)$$

and

$$\Lambda = \bar{\sigma}_{11}^2 - 1. \quad (5.14)$$

The coefficients m_1 and m_3 are formally functions of h and $\bar{\sigma}_{11}$. However, using (4.31b), one can show that dependence on $\bar{\sigma}_{11}$ can be neglected, and substituting $\sigma = \sigma_{11}$ into (5.13) keeps all necessary asymptotic terms. Dependence on σ remains only for the nondimensional frequency parameter Λ by (5.14).

Remark 5.2. *When focusing on the equivalent elliptic orbital tank motions with 'canonic' position of the elliptic trajectory within the framework of Remark 5.1, the right-hand side in (5.12) takes the form $\epsilon_x = \epsilon'_x, \bar{\epsilon}_x = \bar{\epsilon}_y = 0, \epsilon_y = \epsilon'_y$.*

5.2.2 STABILITY

In order to investigate stability of the constructed asymptotic steady-state solution, Lyapunov's linear method and separation of slow and fast time are used. The procedure suggests a slow time scale $\tau = \frac{1}{2}\epsilon^{2/3}t$ and postulates the linearly-perturbed solutions as

$$a_1 = (a + \alpha(\tau))\cos t + (\bar{a} + \bar{\alpha}(\tau))\sin t + o(\epsilon^{1/3}),$$
$$b_1 = (\bar{b} + \bar{\beta}(\tau))\cos t + (b + \beta(\tau))\sin t + o(\epsilon^{1/3}),$$
(5.15)

where $a, \bar{a}, b$, and $\bar{b}$ are defined in (5.9) and computed by (5.12). Substituting (5.15) into the modal equations, grouping the terms of the lowest asymptotic order, and keeping only the linear terms with respect to $\alpha, \bar{\alpha}, \beta$, and $\bar{\beta}$ lead to the linear system of ordinary differential equations

$$\mathbf{s}' + \xi \mathbf{s} + \mathcal{S}\,\mathbf{s} = 0,$$
(5.16)

where $\mathbf{s} = (\alpha, \bar{\alpha}, \beta, \bar{\beta})^T$, the prime means differentiation by τ, and the matrix $\mathcal{S}$ has the following elements

$s_{11} = -2m_1 a\bar{a} - (m_1 - m_3)b\bar{b}; \quad s_{13} = -2m_1\bar{a}b - (m_1 - m_3)a\bar{b},$

$s_{12} = -(\bar{\sigma}_{11}^2 - 1) - m_1(a^2 + 3\bar{a}^2 + b^2) - m_3\bar{b}^2; \quad s_{14} = -2m_3\bar{a}\bar{b} - (m_1 - m_3)ab,$

$s_{21} = (\bar{\sigma}_{11}^2 - 1) + m_1(3a^2 + \bar{a}^2 + \bar{b}^2) + m_3 b^2; \quad s_{22} = 2m_1 a\bar{a} + (m_1 - m_3)b\bar{b},$

$s_{23} = 2m_3 ab + (m_1 - m_3)\bar{a}\bar{b}; \quad s_{24} = 2m_1 a\bar{b} + (m_1 - m_3)\bar{a}b,$

$s_{31} = 2m_1 a\bar{b} + (m_1 - m_3)b\bar{a}; \quad s_{32} = 2m_3 \bar{a}\bar{b} + (m_1 - m_3)ab,$

$s_{33} = 2m_1 b\bar{b} + (m_1 - m_3)a\bar{a}; \quad s_{34} = (\bar{\sigma}_{11}^2 - 1) + m_1(b^2 + 3\bar{b}^2 + a^2) + m_3 \bar{a}^2,$

$s_{41} = -2m_3 ab - (m_1 - m_3)\bar{a}\bar{b}; \quad s_{42} = -2m_1 \bar{a}b - (m_1 - m_3)a\bar{b},$

$s_{43} = -(\bar{\sigma}_{11}^2 - 1) - m_1(3b^2 + \bar{b}^2 + \bar{a}^2) - m_3 a^2; \quad s_{44} = -2m_1 b\bar{b} - (m_1 - m_3)a\bar{a}.$

The fundamental solution $\mathbf{s} = \exp(\lambda\tau)\mathbf{a}$ of the system (5.16) leads to the spectral matrix problem $[(\lambda+\xi)I + \mathcal{S}]\mathbf{a} = 0$, where λ are unknown eigenvalues, $\mathbf{a}$ are the corresponding eigenvectors, and I is the unit matrix. The calculations give the following characteristic (biquadratic) equation

$$(\lambda + \xi)^4 + s_1(\lambda + \xi)^2 + s_0 = 0,$$
(5.17)

where s_0 is the determinant of $\mathcal{S}$. Eigenvalues λ can be defined as $-\xi \pm \sqrt{x_{1,2}}$, where $x_{1,2} = \frac{1}{2}(-s_1 \pm \sqrt{s_1^2 - 4s_0})$ are two solutions of the equation $x^2 + s_1 x + s_0 = 0$.

The asymptotic periodic solution associated with the lowest-order amplitude parameters $a, \bar{a}, b$, and $\bar{b}$, is asymptotically stable ($\alpha, \bar{\alpha}, \beta$, and $\bar{\beta}$ exponentially decrease with increasing τ), if and only if, the real component of λ is strictly negative. For $\xi > 0$, the stability condition can be written as the alternative

either $s_1^2 - 4s_0 \geq 0$ & $-s_1 + \sqrt{s_1^2 - 4s_0} \leq 0$ ($\Leftrightarrow s_0 \geq 0$ & $s_1 \geq 0$), or $s_1^2 - 4s_0 \geq 0$ & $-s_1 + \sqrt{s_1^2 - 4s_0} > 0$ & $\sqrt{\frac{1}{2}\left(-s_1 + \sqrt{s_1^2 - 4s_0}\right)} < \xi$,

or $s_1^2 - 4s_0 < 0$ & $\sqrt{2\sqrt{s_0} - s_1} < \xi$. (5.18)

5.2.3 SUMMARISING THEOREM

Theorem 5.1

Within the framework of the Narimanov-Moiseev asymptotic theory of resonant sloshing in an upright circular base container, which assumes (i) the small-amplitude ($= O(\epsilon) \ll 1$) periodic forcing by surge, sway, roll, and pitch; (ii) the forcing frequency σ is close to the lowest natural sloshing frequency σ_{11} to satisfy the Moiseev detuning ($\Lambda = O(\epsilon^{2/3})$), when Λ is the nondimensional amplitude parameter by (5.14); and (iii) no secondary resonance (the non-dimensional liquid depth $1.05 \lesssim h$ according to [52]), the periodic asymptotic solution of the Narimanov-Moiseev–type modal equations (4.32)–(4.35) is determined by the four lowest-order amplitude parameters $a, \bar{a}, \bar{b}, b$ to be found from the solvability (secular) system (5.12), where $\epsilon_x, \bar{\epsilon}_x, \bar{\epsilon}_y$, and ϵ_y are associated with the lowest Fourier harmonics in the right-hand side of (4.32) [see, (5.3), (5.4)]. Stability of this asymptotic solution follows from the alternative (5.18) and is a function of $a, \bar{a}, \bar{b}, b$, and Λ.

The periodic solution is asymptotically equivalent (have the same first- and second-order components and stability properties) to that occurring for a suitable horizontal orbital elliptic tank motion. When Ox coincides with the major axis of the elliptic tank trajectory, the right-hand side of (5.12) simplifies to $\epsilon_x = \epsilon'_x, \epsilon_y = \epsilon'_y, \bar{\epsilon}_x = \bar{\epsilon}_y = 0$, where the forcing amplitude parameters are computed through the elliptic semi-axes by (5.8). ∎

5.2.4 THE THIRD-ORDER COMPONENTS

Adopting the dominant amplitude parameters $a, \bar{a}, b$ and $\bar{b}$, which are solutions of the secular equations (5.12) and the second-order contribution (5.10) makes it possible to find the asymptotic periodic solution up to the $O(\epsilon)$-order. For the *orbitally-forced container* by (5.5), this solution takes the following form

$$p_{11}(t) = a \cos t + \bar{a} \sin t + \frac{l_3^{(1)}}{9 - \bar{\sigma}_{11}^2}\left[(2\,\bar{a}b\bar{b} + a\,(-a^2 - \bar{b}^2 + b^2 + 3\bar{a}^2))\cos 3t\right.$$
$$\left. + (-2\,ab\bar{b} + \bar{a}\,(\bar{a}^2 + b^2 - \bar{b}^2 - 3a^2))\sin 3t\right] + o(\epsilon), \quad (5.19a)$$

$$r_{11}(t) = \bar{b}\cos t + b\sin t + \frac{l_3^{(1)}}{9 - \bar{\sigma}_{11}^2}\left[(2\,a\bar{a}b + \bar{b}\,(-\bar{b}^2 - a^2 + \bar{a}^2 + 3b^2))\cos 3t\right.$$

$$+ (-2a\bar{a}\bar{b} + b(b^2 + \bar{a}^2 - a^2 - 3\bar{b}^2)) \sin 3t \Big] + o(\epsilon); \quad (5.19b)$$

$$p_{1n}(t) = \frac{1}{1 - \bar{\sigma}_{1n}^2} \Bigg\{ \Bigg[-k_{1n} P_n e_x + \left(l_1^{(n)} - l_2^{(n)} \right) \bar{a} b \bar{b}$$
$$+ a \left(l_1^{(n)} [a^2 + \bar{b}^2 + \bar{a}^2] + l_2^{(n)} b^2 \right) \Bigg] \cos t$$
$$+ \Bigg[-k_{1n} P_n \bar{e}_x + \left(l_1^{(n)} - l_2^{(n)} \right) a b \bar{b} + \bar{a} \left(l_1^{(n)} [\bar{a}^2 + a^2 + b^2] + l_2^{(n)} \bar{b}^2 \right) \Bigg] \sin t \Bigg\}$$
$$+ \frac{l_3^{(n)}}{9 - \bar{\sigma}_{1n}^2} \Bigg\{ \Big[2 \bar{a} b \bar{b} + a \left(-a^2 - \bar{b}^2 + b^2 + 3\bar{a}^2 \right) \Big] \cos 3t$$
$$+ \Big[-2 a b \bar{b} + \bar{a} \left(\bar{a}^2 + b^2 - \bar{b}^2 - 3a^2 \right) \Big] \sin 3t \Bigg\} + o(\epsilon), \quad (5.20a)$$

$$r_{1n}(t) = \frac{1}{1 - \bar{\sigma}_{1n}^2} \Bigg\{ \Bigg[-k_{1n} P_n \bar{e}_y + \left(l_1^{(n)} - l_2^{(n)} \right) a \bar{a} b$$
$$+ \bar{b} \left(l_1^{(n)} [\bar{b}^2 + a^2 + b^2] + l_2^{(n)} \bar{a}^2 \right) \Bigg] \cos t$$
$$+ \Bigg[-k_{1n} P_n e_y + \left(l_1^{(n)} - l_2^{(n)} \right) a \bar{a} \bar{b} + b \left(l_1^{(n)} [b^2 + \bar{b}^2 + \bar{a}^2] + l_2^{(n)} a^2 \right) \Bigg] \sin t \Bigg\}$$
$$+ \frac{l_3^{(n)}}{9 - \bar{\sigma}_{1n}^2} \Bigg\{ \Big[2 a \bar{a} b + \bar{b} \left(-a^2 - \bar{b}^2 + \bar{a}^2 + 3b^2 \right) \Big] \cos 3t$$
$$+ \Big[-2 a \bar{a} \bar{b} + b \left(b^2 + \bar{a}^2 - a^2 - 3\bar{b}^2 \right) \Big] \sin 3t \Bigg\} + o(\epsilon), \quad n \geq 2; \quad (5.20b)$$

$$p_{3n}(t) = \frac{k_1^{(n)}}{1 - \bar{\sigma}_{3n}^2} \Bigg[\left(-2 \bar{a} b \bar{b} + a (a^2 + \bar{a}^2 - b^2 - 3\bar{b}^2) \right) \cos t$$
$$+ \left(-2 a b \bar{b} + \bar{a} (\bar{a}^2 + a^2 - b^2 - 3\bar{b}^2) \right) \sin t \Bigg]$$
$$+ \frac{k_2^{(n)}}{9 - \bar{\sigma}_{3n}^2} \Bigg[\left(6 \bar{a} b \bar{b} + a (a^2 + 3(b^2 - \bar{a}^2 - \bar{b}^2)) \right) \cos 3t$$
$$- \left(6 a \bar{a} b + \bar{a} (\bar{a}^2 + 3(\bar{b}^2 - b^2 - a^2)) \right) \sin 3t \Bigg] + o(\epsilon), \quad (5.21a)$$

$$r_{3n}(t) = \frac{k_1^{(n)}}{1 - \bar{\sigma}_{3n}^2} \Bigg[\left(2 a \bar{a} b + \bar{b} (-\bar{b}^2 - b^2 + \bar{a}^2 + 3a^2) \right) \cos t$$
$$+ \left(2 a \bar{a} \bar{b} + b (-b^2 - \bar{b}^2 + a^2 + 3\bar{a}^2) \right) \sin t \Bigg]$$

$$+ \frac{k_2^{(n)}}{9 - \bar{\sigma}_{3n}^2} \left[-\left(6 a\bar{a}b + \bar{b}(\bar{b}^2 + 3(\bar{a}^2 - a^2 - b^2))\right) \cos 3t \right.$$
$$\left. + \left(6 a\bar{a}\bar{b} + b(b^2 + 3(a^2 - \bar{b}^2 - \bar{a}^2))\right) \sin 3t \right] + o(\epsilon), \quad n \geq 1, \quad (5.21\text{b})$$

where

$$l_1^{(n)} = -\tfrac{3}{4}d_{16,n} + \tfrac{1}{4}d_{18,n} + \sum_{k=1}^{I_r} \left[c_{1k} \left(-\tfrac{1}{2}d_{20,n}^{(k)} - 2d_{21,n}^{(k)} + d_{22,n}^{(k)} \right) \right.$$
$$\left. + s_{1k} \left(-\tfrac{1}{2}d_{23,n}^{(k)} - 2d_{24,n}^{(k)} + d_{25,n}^{(k)} \right) + c_{0k}d_{20,n}^{(k)} - s_{0k}d_{23,n}^{(k)} \right], \quad (5.22\text{a})$$

$$l_2^{(n)} = -\tfrac{1}{4}d_{16,n} + \tfrac{3}{4}d_{18,n} - d_{19,n} + \sum_{k=1}^{I_r} \left[c_{1k} \left(-\tfrac{3}{2}d_{20,n}^{(k)} - 6d_{21,n}^{(k)} + 3d_{22,n}^{(k)} \right) \right.$$
$$\left. + s_{1k} \left(\tfrac{1}{2}d_{23,n}^{(k)} + 2d_{24,n}^{(k)} - d_{25,n}^{(k)} \right) + c_{0k}d_{20,n}^{(k)} - s_{0k}d_{23,n}^{(k)} \right], \quad n \geq 2; \quad (5.22\text{b})$$

$$l_3^{(1)} = \tfrac{1}{2}d_1 + \sum_{k=1}^{I_r} \left[c_{1k} \left(\tfrac{3}{2}d_3^{(k)} + 2d_4^{(k)} \right) + s_{1k} \left(\tfrac{3}{2}d_5^{(k)} + 2d_6^{(k)} \right) \right], \quad (5.22\text{c})$$

$$l_3^{(n)} = \tfrac{1}{4}d_{16,n} + \tfrac{1}{4}d_{18,n} + \sum_{k=1}^{I_r} \left[c_{1k} \left(\tfrac{1}{2}d_{20,n}^{(k)} + 2d_{21,n}^{(k)} + d_{22,n}^{(k)} \right) \right.$$
$$\left. + s_{1k} \left(\tfrac{1}{2}d_{23,n}^{(k)} + 2d_{24,n}^{(k)} + d_{25,n}^{(k)} \right) \right], \quad n \geq 2; \quad (5.22\text{d})$$

$$k_1^{(n)} = -\tfrac{3}{4}d_{11,n} + \tfrac{1}{4}d_{12,n} + \sum_{k=1}^{I_r} \left[c_{1k} \left(-\tfrac{1}{2}d_{13,n}^{(k)} - 2d_{14,n}^{(k)} + d_{15,n}^{(k)} \right) - c_{0k}d_{13,n}^{(k)} \right],$$
$$(5.22\text{e})$$

$$k_2^{(n)} = -\tfrac{1}{4}d_{11,n} - \tfrac{1}{4}d_{12,n} - \sum_{k=1}^{I_r} c_{1k} \left(\tfrac{1}{2}d_{13,n} + 2d_{14,n} + d_{15,n} \right), \quad n \geq 1. \quad (5.22\text{f})$$

In all these expressions, c_{0k}, s_{0k}, c_{1k} and s_{1k} are calculated by using (5.11) and $\bar{\sigma}_{11} = 1$.

Remark 5.3. *As we noted above, the constructed asymptotic nondimensional steady-state solution suggests the orbital horizontal tank motion by (5.5). For the generic periodic forcing by (5.1), this solution should contain an additional linear-type contribution $\tilde{p}_{1n}(t)$ and $\tilde{r}_{1n}(t)$, $n \geq 1$ to the generalised coordinates $p_{1n}(t)$ and $r_{1n}(t)$, $n \geq 1$. To get this extra contribution, one should consider* *the auxiliary forcing*

$$\tilde{\eta}_1(t) = \sum_{k=1}^{\infty} \left[\eta_{1a}^{(k)} \cos kt + \mu_{1a}^{(k)} \sin kt\right] - e_x \cos t - \bar{e}_x \sin t,$$

$$\tilde{\eta}_2(t) = \sum_{k=1}^{\infty} \left[\eta_{2a}^{(k)} \cos kt + \mu_{2a}^{(k)} \sin kt\right] - \bar{e}_y \cos t - e_y \sin t, \qquad (5.23)$$

$$\tilde{\eta}_4(t) = \eta_4(t), \quad \tilde{\eta}_5(t) = \eta_5(t),$$

and use the linearised modal equations (4.32), (4.35) (without the nonlinear quantities $\mathcal{P}_{1n}$ and $\mathcal{Q}_{1n}$, $n \geq 1$), i.e.,

$$\begin{aligned}\ddot{p}_{1n} + \bar{\sigma}_{1n}^2 p_{1n} &= -(\ddot{\tilde{\eta}}_1 - g\tilde{\eta}_5 - S_n\ddot{\tilde{\eta}}_5)\kappa_{1n}P_n, \\ \ddot{r}_{1n} + \bar{\sigma}_{1n}^2 r_{1n} &= -(\ddot{\tilde{\eta}}_2 + g\tilde{\eta}_4 + S_n\ddot{\tilde{\eta}}_4)\kappa_{1n}P_n, \quad 1 \geq 2.\end{aligned} \qquad (5.24)$$

For the generic forcing (5.1), the steady-state solution will operate with the same $p_{0i}(t), p_{2i}(t), r_{2i}(t)$ and $p_{3i}(t), r_{3i}(t)$, $i \geq 1$ by (5.10) and (5.21), but $p_{1n}(t) := p_{1n}(t) + \tilde{p}_{1n}(t)$ and $r_{1n}(t) := r_{1n}(t) + \tilde{r}_{1n}(t)$ by (5.19), (5.20) with solutions of (5.24). The dominant hydrodynamic generalised coordinates p_{11} and r_{11} contain the same lowest Fourier harmonics but may have extra higher Fourier harmonic components. The corrected third-order hydrodynamic generalised coordinates with (5.20) and (5.24) can sufficiently change if the forcing differs from the horizontal elliptic orbit.

5.3 ALTERNATIVE SECULAR EQUATIONS

Remark 5.4. *According to Remark 5.1, appropriate position of the Oxy plane when Ox coincides with the major axis of the elliptic tank orbit and, therefore, the asymptotically equivalent elliptic forcing may be defined by (5.7), simplifies the secular system (5.12), and leads to*

$$\bar{e}_x = \bar{e}_y = 0, \quad \epsilon_x = \epsilon'_x, \quad \epsilon_y = \epsilon'_y \qquad (5.25)$$

by (5.8). This fact was already mentioned in Remark 5.2. Within the framework of the simplification (5.25), one can make useful conclusions on the undamped sloshing and, in addition, transform the 'damped' secular system to a more physically-adequate form.

5.3.1 UNDAMPED SLOSHING

The undamped steady-state wave regimes due to the orbital forcing were analysed in [52]. The analysis suggested the canonic position of the Oxy plane associated with Remarks 5.1 and 5.4, which implies, in particular, $\bar{\epsilon}_x = \bar{\epsilon}_y = 0$. When considering (5.12) with $\xi = 0$, the paper [52] proved that $\bar{a} = \bar{b}$ and, if, in addition, $\epsilon_x \neq \epsilon_y$ (non-circular orbit), $\bar{a} = \bar{b} = 0$. This means that, except for the circular tank orbit, (5.12) transforms to

$$a[\Lambda + m_1 a^2 + m_3 b^2] = \epsilon_x, \quad b[\Lambda + m_1 b^2 + m_3 a^2] = \epsilon_y. \qquad (5.26)$$

The system (5.26) has exact analytical solutions, which will be considered in Chapters 6 (longitudinal excitations) and 7 (elliptic excitations). When $\epsilon_y \neq 0$ (elliptic excitations), only swirling waves are possible according to (3.14). The longitudinal forcing with $\epsilon_y = 0$ may imply swirling or planar (standing) wave.

The situation becomes much more complicated with the *circular orbital* forcing. This forcing implies $\epsilon = \epsilon_x = \epsilon_y$, $\bar{\epsilon}_x = \bar{\epsilon}_y = 0$, but $\bar{a} = \bar{b} = c$ is, generally, non-zero. By taking ①+③ and ①−③, (5.12) is then rewritten in the form

$$\begin{cases} (a+b)[\Lambda + m_1(a^2+b^2) + (3m_1-m_3)c^2 - (m_1-m_3)ab] = 2\epsilon, \\ (a-b)[\Lambda + m_1(a^2+b^2) + (m_1+m_3)c^2 + (m_1-m_3)ab] = 0, \\ c[\Lambda + m_1(a^2+b^2) + (m_1+m_3)c^2 + (m_1-m_3)ab] = 0, \end{cases} \quad (5.27)$$

in which the homogeneous equations contain identical expressions in the square bracket. These expressions are multiplied by $(a-b)$ and c, respectively.

To analyse (5.27), we introduce $a_+ = (a+b)/2$ and $a_- = (a-b)/2$. When both a_- and c equal to zero, the system (5.27) has the following exact solution:

$$\bar{a} = \bar{b} = 0, \; a_+ = a = b, \; \Lambda = \frac{\epsilon}{a_+} - (m_1+m_3)a_+^2. \quad (5.28)$$

According to section 3.1.2.2, this solution determines the swirling wave mode by (3.18), which is characterised by equal longitudinal (along Ox) and transverse amplitudes (along Oy). When either $a_- \neq 0$ or $c \neq 0$, the identical square brackets in (5.27) must be zero. Mathematically, this means that we have only two independent equations in (5.27) which couple three independent amplitude parameters a, b, and c. These two equations can be rewritten in the form

$$\begin{cases} a_+[\Lambda + 4m_1 a_+^2] = -\dfrac{m_1+m_3}{2(m_1-m_3)}\epsilon, \\ a_-^2 + c^2 = -\dfrac{\Lambda + (3m_1-m_3)a_+^2}{(m_1+m_3)} > 0, \; m_1+m_3 > 0, \; m_1 \neq m_3. \end{cases} \quad (5.29)$$

Here, the first equation computes a_+ versus the frequency parameter Λ, but the second one determines the sum $a_-^2 + c^2$ for the given a_+ and Λ. The latter means that the undamped sloshing is characterised by a continuum set of possible steady-state solutions. This contradicts to the existing experimental data. One can show that this contradiction can be resolved when introducing a non-zero damping in the hydrodynamic system.

5.3.2 DAMPED SLOSHING

Following [70], we rewrite (5.12) in a more physically adequate form through introducing the integral amplitudes A, B, and the phase lags ψ, φ defined by

$$A = \sqrt{a^2 + \bar{a}^2} \;\; \text{and} \;\; B = \sqrt{b^2 + \bar{b}^2} > 0, \quad (5.30a)$$

$$a = A\cos\psi, \quad \bar{a} = A\sin\psi, \quad \bar{b} = B\cos\varphi, \quad b = B\sin\varphi. \tag{5.30b}$$

By substituting (5.30) into expressions $\bar{a}①-a②$, $\bar{b}③-b④$, $a①+\bar{a}②$ and $b③+\bar{b}④$ (5.12), we arrive at the following alternative secular equations

$$\begin{cases} \boxed{1}: A[\Lambda + m_1 A^2 + (m_3 - \mathcal{F})B^2] = \epsilon_x \cos\psi, \\ \boxed{2}: B[\Lambda + m_1 B^2 + (m_3 - \mathcal{F})A^2] = \epsilon_y \sin\varphi, \\ \boxed{3}: A[\mathcal{D}B^2 + \xi] = \epsilon_x \sin\psi, \\ \boxed{4}: B[\mathcal{D}A^2 - \xi] = \epsilon_y \cos\varphi, \end{cases} \tag{5.31a}$$

$$\begin{aligned} \mathcal{F} &= (m_3 - m_1)\cos^2(\alpha) = (m_3 - m_1)/(1+C^2), \\ \mathcal{D} &= (m_3 - m_1)\sin(\alpha)\cos(\alpha) = (m_3 - m_1)C/(1+C^2), \end{aligned} \tag{5.31b}$$

where

$$\Lambda = \bar{\sigma}_{11}^2 - 1, \quad \alpha = \varphi - \psi, \quad C = \tan\alpha, \quad 0 \leq \epsilon_y \leq \epsilon_x \neq 0,$$

($\mathcal{F}(\alpha)$ and $\mathcal{D}(\alpha)$ are the π-periodic functions by the phase lag difference α).

Secular systems (5.12) and (5.31) are *mathematically equivalent*. In other words, having known A, B, ψ, φ from (5.31), we can define $a, \bar{a}, b, \bar{b}$ and vice versa. Chapters 6 and 7 will utilize (5.30) in analytical and numerical studies of the steady-state sloshing.

Remark 5.5. *According to the Narimanov-Moiseev asymptotic theory, expression (3.18) determines the lowest-order approximation of the resonant steady-state modes when replacing $\sigma_{11} \to \sigma$. This means that the nondimensional asymptotic approximation of the free-surface elevation is defined by*

$$\zeta(r,\theta,t) = \alpha_{11}J_1(k_{11}r)[(a\cos\theta + \bar{b}\sin\theta)\cos t + (\bar{a}\cos\theta + b\sin\theta)\sin t] + o(\epsilon^{2/3}). \tag{5.32}$$

As discussed in section 3.1.2.2, expression (5.32) describes either swirling or standing wave mode, and the criterion (3.14) holds true to discriminate these wave modes. In terms of the introduced notations (5.30) and (5.31), the criterion (3.14) implies the alternative ($\Xi = ab - \bar{a}\bar{b}$):

$$\begin{aligned} &\Xi, \ \sin\alpha > 0 \ (counterclockwise \ swirling), \\ &\Xi, \ \sin\alpha, \ C = 0 \ (standing \ wave), \\ &\Xi, \ \sin\alpha < 0 \ (clockwise \ swirling). \end{aligned} \tag{5.33}$$

This criterion will furthermore be used when analysing the resonant sloshing due to the orbital elliptic tank forcing.

6 Steady-state resonant sloshing due to the longitudinal forcing

By adopting the already-introduced notations and definitions as well as using the theoretical results on the periodic (steady-state) solution of the Narimanov-Moiseev–type equations, this chapter concentrates on the particular case of combined surge-and-pitch periodic resonant tank excitations (i.e., $\eta_1(t) \sim \eta_5(t) = O(\epsilon)$, $\eta_2(t) = \eta_4(t) \equiv 0$ in Figure 4.1). Sloshing due to this kind of forcing is a typical benchmark problem. Systematic experimental and theoretical studies of the problem were initiated in the NASA programs of the 1950s and 1960s [2]. Pioneering theoretical analysis of the corresponding steady-state waves can be found in the papers by John Miles [161–163] and Ivan Lukovsky [138, 140, 143, 147]. Recently, the problem has, in detail, been revisited in the experimental work [203]. Emil Hophinger received the EUROMECH award for these experimental investigations.

From all existing theoretical and experimental works, we know that the longitudinal resonant forcing leads to planar and/or swirling steady-state waves; moreover, there are the forcing frequencies for which these wave modes are not stable, and therefore, irregular (chaotic, modulated, etc.) sloshing occurs. The resonant sloshing problem was theoretically analysed in [52] by using the Narimanov-Moiseev–type modal system from Chapter 4 with the zero damping, but [198] considered the damped steady-state sloshing. In the current chapter, we follow these papers. For both the undamped and damped cases, we construct analytical solutions of the secular system (5.12) and, thereby, derive exact analytical asymptotic (steady-state) solutions of the Narimanov-Moiseev–type modal system. These solutions are classified by drawing the wave-amplitude response curves.

6.1 STEADY-STATE WAVE MODES

6.1.1 UNDAMPED SLOSHING

Within the framework of nonlinear asymptotic analysis of the Narimanov-Moiseev–type system from the previous chapters, the undamped resonant sloshing due to the longitudinal forcing causes the same secular system (5.12) with respect to the four nondimensional amplitude parameters $a, \bar{a}, \bar{b}$, and b, in which $\bar{\epsilon}_x = \epsilon_y = \bar{\epsilon}_y = 0$ and $\xi = 0$. Rigorous mathematical analysis in [52] proved that the necessary solvability condition of the secular system reads then as $\bar{a} = \bar{b} = 0$. When remembering (5.30b), this fact means that the identity $\sin\psi = 0 = \cos\varphi$ is fulfilled for the undamped longitudinally excited resonant sloshing. Furthermore, [52] shows that the secular system (5.12) has two and only two solution types, which determine planar (standing) waves [$a \neq 0$, $\bar{a} = \bar{b} = b = 0$, $\sin\psi = 0$, φ is not defined] and swirling [$ab \neq 0$, $\bar{a} = \bar{b} = 0$, $\sin\psi = \cos\varphi = 0$]. The planar waves (solutions) are associated with real roots of the cubic equation

$$a[\Lambda + m_1 a^2] = \epsilon_x, \tag{6.1}$$

but the two non-zero swirling wave-amplitude parameters a and b come from

$$a[\Lambda + (m_1 + m_3)a^2] = \frac{m_1}{m_1 - m_3}\epsilon_x, \quad b^2 = -\frac{\Lambda + m_3 a^2}{m_1} > 0, \tag{6.2}$$

where the (first) cubic equation determines a versus the forcing frequency parameter Λ, but the second formula computes b^2 for the given Λ and a. Reformulating this result in terms of the secular system (5.31) leads to the following theorem.

Theorem 6.1: Undamped resonant sloshing

There exist only two undamped solution types of the secular system (5.31) for the longitudinal harmonic forcing along the Ox-axis ($\xi = \bar{\epsilon}_x = \bar{\epsilon}_y = \epsilon_y = 0$). They imply the planar standing ($A > 0, B = 0, \sin\psi = 0, C = 0$, phase lag φ is not defined) and swirling ($A > 0, B > 0, \sin\psi = \cos\varphi = 0, C = \pm\infty$) wave modes. When all the steady-state waves (solutions) are unstable, sloshing possesses an irregular character (chaotic, modulated, etc. waves).

For each fixed pair $A, B > 0$ of the wave-amplitude parameters, the corresponding swirling (angularly-propagating) wave may occur in either counterclockwise or clockwise direction, which is determined by the phase lag difference $\alpha = \varphi - \psi$ (alternatively, by the sign of C): $\alpha = \pi/2$ ($C = +\infty$) means the counterclockwise wave propagation and $\alpha = -\pi/2$ ($C = -\infty$) corresponds to the clockwise swirling. ∎

6.1.2 DAMPED SLOSHING

This book mainly deals with the strictly positive damping coefficient $\xi > 0$. The authors [198] analysed this case by using the secular system (5.31) and proved the following theorem.

Theorem 6.2: Damped resonant sloshing

For the longitudinal tank forcing along the Ox axis ($\epsilon_x > 0$, $\bar{\epsilon}_x = \bar{\epsilon}_y = \epsilon_y = 0$), a small non-zero damping ($\xi > 0$) in the secular system (5.31) causes two and only two solution types implying the two steady-state wave modes, i.e., planar standing and swirling. Specifically,

(A) The planar standing wave corresponds to solutions of (5.31), with $B = 0$, $A > 0$, and $C = 0$ (phase lag φ is not defined), where A and ψ are analytically determined by the equations

$$A^2\left[(\Lambda + m_1 A^2)^2 + \xi^2\right] = \epsilon_x^2; \quad 0 < A \leq \frac{\epsilon_x}{\xi}; \quad \psi = \arccos \frac{A(\Lambda + m_1 A^2)}{\epsilon_x}. \quad (6.3)$$

(B) The steady-state swirling waves correspond to solutions of (5.31) with $A > 0, B > 0, C \neq 0$, which can be sequentially found by solving the cubic equation

$$q_3 C^3 + q_2 C^2 + q_1 C + q_0 = 0, \quad (6.4)$$

where

$$q_3 = \xi^3 (m_1 + m_3)^2 > 0, \quad q_2 = 2\xi^2 \Lambda (m_3^2 - m_1^2),$$
$$q_1 = \xi\left[4\xi^2 m_1^2 + \Lambda^2 (m_1 - m_3)^2\right], \quad q_0 = \epsilon_x^2 m_1^2 (m_1 - m_3),$$

and substituting its real roots into expressions

$$A^2 = \frac{\xi(1+C^2)}{(m_3 - m_1)C} > 0, \quad (6.5)$$

and

$$B^2 = -\frac{1}{m_1}\left[\Lambda + \frac{m_1 + m_3}{1 + C^2} C^2 A^2\right] > 0 \quad (6.6)$$

to specify the wave-amplitude parameters A and B. For each fixed $A > 0$ and $B > 0$ by (6.5) and (6.6), there are two identical swirling waves occurring in either counterclockwise or clockwise direction.

(C) When solutions (A) and (B) are unstable, irregular (chaotic, modulated, etc.) waves occur. The stability analysis may adopt the criterion (5.18).

Proof. Let $\xi > 0$ in (5.31) and consider an obvious solution of (5.31) with $B = 0$, which corresponds to the planar standing wave mode. Equations $\boxed{2}$ and $\boxed{4}$ become then identities for any φ, hence, this phase lag is not defined anymore for this wave mode within the framework of the Narimanov-Moiseev theory. Two other equations of (5.31), $\boxed{1}$ and $\boxed{3}$, take then the following form

$$A[\Lambda + m_1 A^2] = \epsilon_x \cos\psi, \quad A\xi = \epsilon_x \sin\psi. \quad (6.7)$$

The sum $\boxed{1}^2 + \boxed{3}^2$ deduces the cubic equation in (6.3) with respect to A^2, which makes it possible to find A^2 as a function of Λ (the forcing frequency parameter). The second equality of (6.3) follows from the first equation in

(6.7). The second equation in (6.7) with $A > 0$, $\xi > 0$, and $\epsilon_x > 0$ deduces $\sin \psi > 0$. This inequality justifies the usage of the 'arccos' function in (6.3). Finally, the upper limit in the middle equation of (6.3) is associated with the case $\sin \psi = 1$. Thus, the statement (A) is fully proven.

Let us now consider the case $\epsilon_y = 0$, $\xi > 0$ and $A > 0, B > 0$. Equation $\boxed{4}$ of (5.31) leads then to the identity $\mathcal{D}A^2 = \xi$, where, accounting for the explicit expression of $\mathcal{D}$ in (5.31b), we get (6.5) and conclude the inequality $C > 0$. By substituting the latter expression into $\boxed{2}$ of (5.31) with $B > 0$ and $\epsilon_y = 0$, we deduce (6.6).

The next step is related to $\boxed{1}$ and $\boxed{3}$ in (5.31), which we rewrite in the form

$$A\left[\Lambda + m_1 A^2 + \frac{m_1 + m_3 C^2}{1 + C^2}B^2\right] = \epsilon_x \cos \psi, \qquad (6.8)$$

$$A\left[\frac{(m_3 - m_1)C}{1 + C^2}B^2 + \xi\right] = \epsilon_x \sin \psi. \qquad (6.9)$$

Taking $(6.8)^2 + (6.9)^2$, in which we sequentially substitute expressions (6.6) and (6.5), we obtain an equation with respect to the parameter $C = \tan \alpha$. Tedious derivatives show that it is exactly the cubic equation (6.4). Thus, the statement (B) is proven. Assertion (C) is obvious. ∎

Remark 6.1. *Theorems 6.1 and 6.2 are proven for the purely harmonic longitudinal forcing. Theorem 5.1 states that these results keep true for any periodic sway/roll/switch/pitch when only $\eta_1(t)$ and $\eta_5(t)$ contain the nonzero first Fourier harmonics in (5.1). An astonishing example is the horizontal tank motion along the Bernoulli lemniscate of the nondimensional width l_a by $\eta_1(t) = l_a \cos t/(1 + \sin^2 t) = 2l_a(\sqrt{2} - 1) \cos t + \ldots$, $\eta_2(t) = l_a \sin t \cos t/(1 + \sin^2 t) = 2l_a(3 - 2\sqrt{2}) \sin 2t + \ldots$. According to the theorem, the lemniscate is asymptotically equivalent to the longitudinal excitation by $\eta_1(t) = 2l_a(\sqrt{2} - 1) \cos t$. The difference between the steady-state sloshing due to original (lemniscate) and equivalent longitudinal trajectories is the $O(\epsilon)$-order wave component, which, in addition, does not affect the hydrodynamic stability.*

6.2 RESPONSE CURVES

The formulas in Theorems 6.1 and 6.2 facilitate a constructive steady-state sloshing analysis for the longitudinal forcing. As a result, one can draw the wave-amplitude and phase-lag response curves on which stable/unstable (section 5.2.2) steady-state solutions are specified. Properties of these response curves depend on coefficients m_1 and m_3, which are functions of the nondimensional liquid depth h. The Narimanov-Moiseev theory is applicable for $1.05 \lesssim h$. Table 8.3 proves the following inequalities for these mean liquid depths

$$O(1) = m_1 < 1, \ O(1) = m_3 > 0 \text{ and } O(1) = m_3 + m_1 > 0. \qquad (6.10)$$

6.2.1 WAVE AMPLITUDES

If coefficients m_1 and m_3 satisfy the inequalities (6.10), the undamped/damped wave-amplitude response curves in the $(\sigma/\sigma_{11}, A, B)$-space have geometric shapes, which are schematically illustrated in Figure 6.1 for $h = 1.5$, the nondimensional forcing amplitude $\eta_{1a} = 0.01$, and the damping coefficient $\xi = 2\xi_{11} = 0.02$. The latter input values are relevant to experiments in [203] with circular cylindrical tanks filled by tap water, whose radii belong to the range $0.05 \text{ m} \leq R_0 \leq 0.3 \text{ m}$. The damping coefficient ξ will be discussed, in context of the theoretical estimates from section 3.3 and actual measurements of the phase lag ψ made in experiments [203].

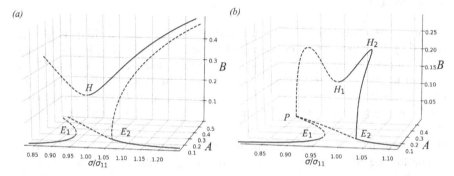

Figure 6.1 The wave-amplitude response curves in the $(\sigma/\sigma_{11}, A, B)$-space for the longitudinal horizontal harmonic forcing. The nondimensional liquid depth $h = 1.5$; the nondimensional forcing amplitude is $\eta_{1a} = 0.01$ ($\eta_{2a} = 0$). The solid lines mark the stable solutions (steady-state waves), but the dashed (deep-blue) lines imply the instability. The undamped case ($\xi = 0$) is presented on the panel (a), and the damped one ($\xi = 0.02$) is in the panel (b). There is no stable steady-state sloshing between E_1 and $H_1(H)$ where irregular (chaotic, modulated, etc.) waves are expected. The curves belonging to the $(\sigma/\sigma_{11}, A)$-plane are responsible for the planar (standing) waves. The non-zero damping causes bifurcation point P where the swirling wave mode emerges from the planar one. Bifurcation point H_2 determines the upper bound (between H_1 and H_2) of the frequency range, where the swirling wave mode is stable.

The response curves belonging to the $(\sigma/\sigma_{11}, A)$-plane are responsible for planar standing waves. The spatial curves correspond to the swirling wave mode. The solid lines indicate stable steady-state waves, but the dashed (deep-blue) lines imply the hydrodynamic instability. The stable planar waves are expected in the left of E_1 and in the right of E_2. They are unstable between E_1 and E_2, where stable swirling (in the right of H (H_1)) and irregular (chaotic, modulated, etc.) waves (that is, there are no stable steady-state waves) are theoretically predicted. Numerical simulations and experimental observations in the latter resonance zone are discussed, for example, by Miles [163, 164] and Ibrahim [100]. When $\xi = 0$, the swirling wave mode always exists for the forcing frequencies larger than the abscissa of H. A novelty of the damped steady-state sloshing consists of two bifurcation points, H_2 and P. The wave-amplitude response curves, which correspond to swirling, constitute the arc clamped at P and E_1. Each point on this arc implies two identical swirling waves occurring in either clockwise or counterclockwise direction. Damping determines the upper frequency bound (H_2) for the swirling wave mode. Points P and H_2 go to infinity as $\xi \to 0$.

6.2.2 PHASE LAG

According to Theorem 6.1, the undamped steady-state sloshing is characterised by the piecewise values of ψ and φ, that is, $\sin \psi = 0$ (φ is not defined) for planar waves and $\sin \psi = \cos \varphi = 0$ for swirling. The result should be applied to the wave-amplitude response curves in Figure 6.1(a), i.e., $\psi = 0$ on the branch with E_1 and $\psi = \pi$ for the branch with E_2, but two different solutions are possible for the H-containing branch (where swirling is stable) with $\varphi = \pi/2$ and $-\pi/2$, and $\psi = 0$.

Theorem 6.2 states that the phase lags ψ and φ are complex functions of the forcing frequency σ/σ_{11} as the damping coefficient $\xi > 0$, i.e., these phase lags continuously vary along the response curves in Figure 6.1(b). The planar waves correspond to (6.3) and swirling is associated with $C = \tan \alpha = \tan(\varphi - \psi) \geq 0$ (the inequality follows from (6.6) where $m_3 > m_1$ according to (6.10)). Positive real roots of the cubic equation (6.4), after being substituted into (6.5) and (6.6), compute A^2, B^2 but not ψ and φ. In order to get ψ, one should insert C and, further, A and B into $\boxed{1}$ and $\boxed{3}$ of (5.31a). There exists a unique ψ for each point of the response curves in Figure 6.1(b). However, there is no unique φ but two, $\varphi_1 = \psi + \alpha$ and $\varphi_2 = \psi + \alpha \pm \pi$, for swirling on $PH_1H_2E_2$. These φ_1 and φ_2 imply two physically-identical swirling waves but occurring in counterclockwise and clockwise directions, respectively.

Adopting notations from Figure 6.1(b), Figure 6.2 demonstrates the theoretical phase-lag response curves for ψ in the $(\sigma_{11}, \psi/\pi)$-plane. It confirms that the phase lag ψ is not a piecewise function as $\xi > 0$. The phase-lag response curves clearly identify the frequency ranges of irregular (chaotic, modulated, etc.) waves between E_1 and H_1, further, only stable swirling waves are realised between H_1 and E_2, but both planar and swirling stable waves are possible between E_1 and H_2. One should recall that two physically-identical stable swirling waves (of different azimuthal directions) are expected on the frequency interval determined by H_1 and H_2. This fact is marked by the corresponding symbols. The double-arrow symbol is used to specify the stable planar standing wave mode. Point H_2 runs to the right with decreasing ξ. This enlarges zone H_1H_2 where stable planar (standing) and swirling waves co-exist.

Steady-state resonant sloshing due to the longitudinal forcing

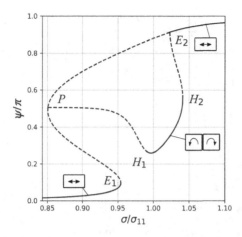

Figure 6.2 Typical phase-lag response curves in the $(\sigma/\sigma_{11}, \psi/\pi)$-plane of the damped steady-state sloshing for the longitudinal forcing, which are associated with the input data in Figure 6.1(b). The figure adopts analogous notations for bifurcation points. Specific symbols are used to mark stable swirling (clockwise and counterclockwise) on $H_1 H_2$ and planar standing waves.

6.2.3 EFFECTIVE FREQUENCY DOMAINS OF STEADY-STATE WAVE MODES

According to the response curves in Figures 6.1(b) and 6.2, effective (stability) frequency ranges of the steady-state wave modes and irregular (chaotic, modulated, etc.) sloshing are determined by abscissas of bifurcation points E_1, E_2, H_1 and H_2. Positions of these points are functions of the forcing amplitude η_{1a} and the damping coefficient ξ (for the fixed liquid depth h).

By assuming the mean liquid depth $h = 1.5$, Figure 6.3 illustrates dependence of the abscissas of the aforementioned bifurcation points *versus* the forcing amplitude η_{2a} and four different damping coefficients in the panels (a) $\xi = 0.0$, (b) $\xi = 0.05$, (c) $\xi = 0.0125$, and (d) $\xi = 0.02$. In these figures, the stable planar wave mode is expected in the left of the line E_1 and in the right of E_2, the stable swirling is between the lines H_1 and H_2, but irregular (chaotic, modulated, etc.) sloshing is predicted between E_1 and H_1.

Figure 6.3(a) corresponds to the undamped resonant sloshing when H_2 is at the infinity. This panel contains both the theoretical expectations by the Narimanov-Moiseev steady-state theory and the corresponding experimental results on the bifurcation point positions in [203]. The results are in a good agreement. The next panels are drawn for the non-zero damping.

Figure 6.3(b) represents calculations done with $\xi = 0.005$. Bearing in mind experiments in [203], this damping coefficient assumes that dissipation in the hydrodynamic system is fully associated with the laminar boundary layer and the bulk viscosity, $0.005 = 2\xi_{11}^{(0)} = 2(\xi_{11}^{surf} + \xi_{11}^{bulk})$, where ξ_{11}^{surf} and ξ_{Mi}^{bulk} are computed by (3.78) and (3.81), respectively. We see that the damping

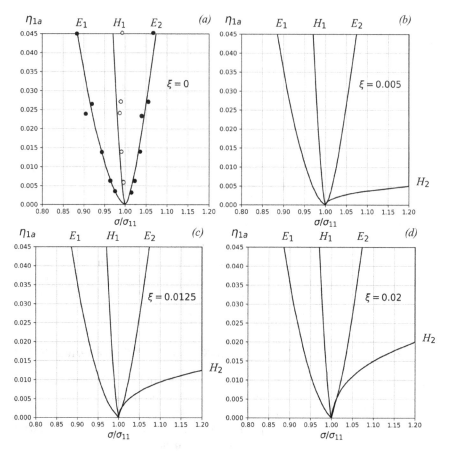

Figure 6.3 Abscissas σ/σ_{11} of the bifurcation points E_1, E_2, H_1, and H_2 versus the nondimensional forcing amplitude η_{1a} and four different damping coefficients. The panel (a) is associated with the undamped sloshing, (b) corresponds to $\xi = 2\xi_{11}^{(0)} = 2(\xi_{11}^{surf} + \xi_{11}^{bulk})$ computed by (3.78) and (3.81), (c) uses the mean averaged value of ξ computed in the next sections from experimental data in [203], and (d) adopts the trial value in Figure 6.1. Stable planar waves are predicted in the left of the line E_1 and in the right of E_2. Stable swirling is between H_1 and H_2, but irregular sloshing occurs between E_1 and H_1. Symbols (circles) in panel (a) come from experimental measurements in [203].

effect on H_2 is only important as $\eta_{1a} \lesssim \xi$. Formally, the panels (c) and (d) with $\xi = 0.0125$ and 0.02 confirm the inequality. However, because $\xi \sim \delta_b$ in (3.73), condition $\eta_{1a} \lesssim \xi$ appearing in the case (b) means that, if the damping is uniquely associated with (3.78) and (3.81), it only matters when the forcing amplitude does not exceed the boundary-layer thickness. The latter condition doesn't have any physical sense – for those small forcing amplitudes, one should consider the viscous hydrodynamic model based on the Navier-Stokes equation. The panels (c) and (d) try larger ξ than within the framework of the boundary layer on the tank walls and the bulk viscosity. Specifically, $\xi = 0.0125$ in the panel (c) corresponds to the averaged integral damping coefficients coming from measurements [203] (this estimate will be made in the next section), but $\xi = 0.02$ is the trial value in the illustrative examples of Figures 6.1 and 6.2. The larger ξ is justified by diverse additional dissipative phenomena.

In *summary*, the damping weakly affects abscissas of bifurcation points E_1, E_2, and H_1 and, therefore, frequency ranges for stable planar waves and irregular sloshing are almost not depending on the integral damping in the hydrodynamic system within the framework of the Narimanov-Moiseev theory. This explains why the undamped theoretical results are well supported by experiments from [203]. The damping matters for the upper bound of the frequency range responsible for the stable swirling waves. Because the adopted hydrodynamic model requires the forcing amplitude η_{1a} be larger than the viscous boundary-layer thickness δ_b, the stability range of the steady-state swirling is only affected by the non-zero integral damping coefficient ξ when the nonlinear free-surface phenomena sufficiently contribute to the global energy dissipation, i.e., $2(\xi_{11}^{surf} + \xi_{11}^{bulk}) \lesssim \xi$.

6.3 COMPARISON WITH EXPERIMENTS

Section 6.2.3 showed that the Narimanov-Moisev theory accurately predicts the experimentally established zones of the stable planar and irregular waves; these zones are weakly affected by the damping. The same is true for the lower-frequency bound of the swirling wave mode; the damping may only influence its upped bound. The literature contains measurements of wave elevations and hydrodynamic forces, which will be compared in the present section with their theoretical values to study how the phase lag, elevations, and forces are affected by the damping.

6.3.1 PHASE LAGS

Figure 6.4 compares measurements of the phase lag ψ and their theoretical predictions for the swirling wave mode (the frequency range between H_1 and H_2 in Figures 6.1(b) and 6.2). The input parameters are associated with [203], where the tank radius is 15 cm, the nondimensional liquid depth $h = 1.5$, liquid is tap water, and the nondimensional forcing amplitude $\eta_{1a} = 0.045$. This provides the Bond number Bo $\simeq 500$, i.e., the case can be studied by using the constructed analytical theory (according to section 3.3).

The viscous laminar boundary layer on the wetted tank surface and the bulk damping predict, according to (3.74), (3.78), and (3.81), the damping coefficient $\xi = \xi_1 = 0.005$. As discussed before, this is the lower bound of the actual ξ, that is, the actual damping coefficient (in experiments) should

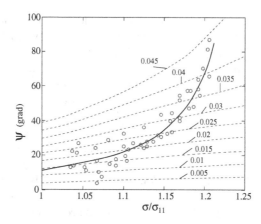

Figure 6.4 The measured phase lag ψ (red circles, degrees) from [203] for the stable swirling wave mode (corresponds to $H_1 H_2$ in Figures 6.1(b) and 6.2) and their theoretical estimates (the dashed lines) with the damping coefficients $\xi_1 \leqslant \xi = \xi_i \leqslant \xi_9$, when $\xi_1 = 0.005$ and $\xi_9 = 0.045$; $h = 1.5$. The longitudinal horizontal harmonic forcing with $\eta_{1a} = 0.045$. The comparison shows that the damping increases with the wave amplitude (along $H_1 H_2$) in the experiments, which may be explained by additional damping factors. The solid line implies the mean-square approximation of the experimental data. The mean value of the experimental damping coefficient is equal to 0.0125.

be larger than ξ_1 since we want to include other dissipative factors in the hydrodynamic system. Figure 6.4 depicts the theoretical phase-lag response curves (the dashed lines) and the measured ψ (circles). The theoretical curves are given for different (but fixed along the curve) values of ξ in the range from $\xi = \xi_1 = 0.005$ to $\xi = \xi_9 = 0.045$.

According to the comparative results in Figure 6.4, $\xi = \xi_1$ indeed provides an estimate from below the experimental phase-lag response curve (the solid line), which was drawn by using the mean-square approximation. To get a consistency between the theory and experiments, one must postulate that the integral damping in the mechanical system (associated with ξ) increases with increasing the wave amplitude along this phase-lag response curve (from H_1 to H_2 in Figure 6.2). As discussed in [203], increasing ξ should be clarified by diverse dissipative free-surface phenomena, which are exemplified by fragmentation of the free surface, wave breaking, etc. The average (mean) value of ξ is estimated about 0.0125. However, ξ, no doubt, is larger when approaching H_2. In *summary*, the damping is very important for the phase lag ψ. In order to fit the experimental phase-lag response curves for the swirling wave mode, the theory should assume that the damping coefficient ξ increases with increasing the wave amplitude.

6.3.2 WAVE ELEVATIONS NEAR THE WALL AND HYDRODYNAMIC FORCE

The maximum free-surface elevations at the distance $0.125R_0$ from the wall (to avoid the capillary meniscus effect) were measured in [203]. By adopting the averaged (mean) damping coefficient $\xi = 0.0125$ from the previous section, we computed (within the framework of the third-order approximation in section 5.2.4 and the free-surface representation (4.11a)) the maximum wave elevation of the steady-state waves at the measured probe. Figure 6.5 compares the computed maximum elevation and measurements from [203]. These are in satisfactory agreement. Numerical experiments with smaller ξ showed that the wave elevations are weakly affected by the damping, and one can use the undamped sloshing predictions for approximating the wave elevations. Another problem is the upper bound of the forcing frequency associated with H_2 (for swirling). Section 6.2.3 showed that position of H_2 is sensitive to ξ, in general, and may be affected by the free-surface phenomena like the wave breaking. As explained in the context of Figure 6.3, increasing η_{1a} makes the integral damping less important.

Figure 6.6 depicts the theoretical steady-state nondimensional horizontal hydrodynamic force $F_1(t)$ (see section 4.4.1) and the average (mean) amplitude defined as one-half of the wave height of the nondimensional wave elevations near the wall for the planar wave mode. These force and amplitude parameters were measured in [2]; these experimental values are marked by symbols. Abramson [2] used different tanks with R_0 between 8 and 18 cm and

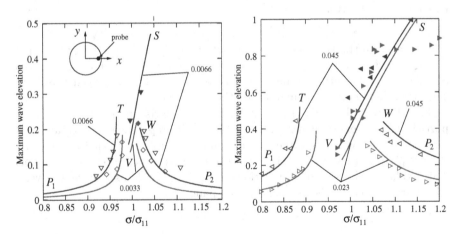

Figure 6.5 The theoretical (lines) and experimental (symbols) nondimensional maximum wave elevations measured at the point $(0.875\,R_0, 0)$. The translational tank forcing with nondimensional amplitudes $\eta_{1a} = 0.0033, 0.0066, 0.023$, and 0.045. The calculations were done by adopting the damping coefficient $\xi = 0.0125$. The empty symbols correspond to the planar standing waves, and the filled ones imply the swirling wave mode.

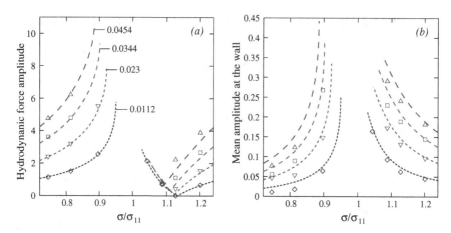

Figure 6.6 Theoretical and measured steady-state nondimensional horizontal hydrodynamic force and 'average' amplitude of elevations at the wall versus the forcing frequency for the planar wave regime. The measurements are taken from [2]; $h = 2$ and different radii R_0 were in the experimental setup from [2]. The theoretical curves are marked by the nondimensional horizontal forcing amplitudes η_{1a}. The experimental symbols are: $\Diamond - \eta_{1a} = 0.0112$, $\triangledown - 0.023$, $\square - 0.0344$, and $\triangle - 0.0454$. The nondimensional horizontal hydrodynamic force amplitude is computed as $\max_{0 \leqslant t < 2\pi} 100|F_1(t)|/(\rho \bar{g})(2R_0)^3$, where $\bar{g}$ is the dimensional gravity acceleration. The average amplitude at the wall means one-half of the wave height near the wall. The distance from the wall was not specified, and the theoretical values are computed at $(x, y) = (1, 0)$.

$h = 2$ for which the Bond number is about 1000. However, the lower bound of $\xi = 2\xi_{11}^{low}$ dramatically changes from tank to tank (see Figure 3.5). To avoid conflicts, the computations have been done for $\xi = 0$ (undamped sloshing). The theoretical curves are marked by the η_{1a}-values.

The theoretical steady-state results for the hydrodynamic force are supported by these experiments (the panel a). A discrepancy is only established for $\sigma/\sigma_{11} > 1.1$ with the larger forcing amplitudes $\eta_{1a} = 0.0454$ and 0.0344. Report [2] did not specify positions of the measurement probe for elevations. This explains why we computed the average amplitude exactly at the wall. This partly explains discrepancies in the panel (b).

7 Steady-state sloshing for the orbital forcing

7.1 WAVE-AMPLITUDE PARAMETERS

We consider steady-state periodic resonant waves due to orbital cyclic sway/role/pitch/surge motions of an upright circular tank with the forcing frequency close to the lowest natural sloshing frequency and a relatively small forcing amplitude. According to Theorem 5.1, periodic asymptotic solutions of the Narimanov-Moiseev–type modal equations, which describe these waves, are asymptotically equivalent to solutions occurring due to a horizontal elliptic forcing. Without loss of generality, the analysis can assume that the elliptic tank trajectory has the major axis along Ox so that one can adopt conclusions from Remark 5.4 and utilise the alternative secular system in section 7.2.2. Concentrating on the elliptical trajectories means that (5.8) is fulfilled in (5.12) (or (5.31)). The latter implies $\delta_1 = \epsilon_y/\epsilon_x \neq 0$, where $|\delta_1|$ is the semi-axes ratio.

As we have seen in the previous chapters, when $\xi > 0$, the phase lags φ and ψ are complex functions of the input parameters. To remove φ and ψ and simplify the secular systems (5.31a) for the elliptic forcing by (5.7) and (5.8), we substitute $\varphi = \psi + \alpha$ into the right sides $\boxed{2}$ and $\boxed{4}$ and take $\epsilon_x \cos\psi$ and $\epsilon_x \sin\psi$ from $\boxed{1}$ and $\boxed{3}$. The result is the following linear system of homogeneous equations

$$\begin{cases} (\delta_1 A)[\cos\alpha(\mathcal{D}B^2 + \xi) + \sin\alpha(\Lambda + m_1 A^2 + (m_3 - \mathcal{F})B^2)] \\ \qquad\qquad - B[\Lambda + m_1 B^2 + (m_3 - \mathcal{F})A^2] = 0, \\ (\delta_1 A)[\cos\alpha(\Lambda + m_1 A^2 + (m_3 - \mathcal{F})B^2) - \sin\alpha(\mathcal{D}B^2 + \xi)] \\ \qquad\qquad - B[\mathcal{D}A^2 - \xi] = 0, \end{cases}$$

with respect to two non-zero variables $\delta_1 A$ and B. The system should have a non-trivial solution. This leads to the solvability (zero-determinant condition)

$$\xi(A^2 - B^2)(m_3 - m_1)\sin\alpha - \cos\alpha[\xi^2 + (\Lambda + m_1(A^2 + B^2))^2] \\ + (m_3 - m_1)^2 A^2 B^2 \cos^2\alpha \sin^2\alpha = 0, \quad (7.1)$$

which *does not depend* on δ_1. Other two equations with respect to A^2, B^2 and C follow from $\boxed{1}^2 + \boxed{3}^2$ and $\boxed{2}^2 + \boxed{4}^2$:

$$\begin{cases} A^2[(\Lambda + m_1 A^2 + (m_3 - \mathcal{F})B^2)^2 + (\mathcal{D}B^2 + \xi)^2] = \epsilon_x^2, \\ B^2[(\Lambda + m_1 B^2 + (m_3 - \mathcal{F})A^2)^2 + (\mathcal{D}A^2 - \xi)^2] = \delta_1^2 \epsilon_x^2; \end{cases} \quad (7.2)$$

they do not depend on the sign of δ_1 (direction of the orbital forcing).

The system (7.1), (7.2) possesses the following properties.

Assertion 7.1. *For the elliptic forcing with $\delta_1 \neq 0$, the secular system (7.1), (7.2) does not depend on the sign of δ_1 (direction of the elliptic orbital excitation), therefore, the wave-amplitude response curves in the $(\sigma/\sigma_1, A, B)$-space do not depend on the forcing direction; swirling waves should, therefore, be interpreted as either co- or counter-directed to the tank orbit.*□

Assertion 7.2. *Standing wave mode is not possible.*

Proof. Indeed, the standing steady-state wave mode corresponds to solutions of (7.1), (7.2) with $C = 0$ ($\sin \alpha = 0$). Assume that $C = 0$. This leads to (7.1) $\Rightarrow \xi^2 + (\Lambda + m_1(A^2 + B^2))^2 = 0$ and (7.2) $\Rightarrow A^2[\xi^2 + (\Lambda + m_1(A^2 + B^2))^2] = \epsilon_x^2 \neq 0$, which is impossible. □

Assertion 7.3. *The phase lag difference $\alpha = \varphi - \psi$ cannot be the root of $\cos \alpha = 0$ as $0 < \delta_1 < 1$.*

Proof. When $\cos \alpha = 0$, equation (7.1) leads to $A^2 = B^2$, $F = D = 0$ and, therefore, (7.2) gives

$$\epsilon_x^2 = A^2[(\Lambda + (m_1 + m_3)A^2)^2 + \xi^2] = \delta_1^2 \epsilon_x^2 \neq \epsilon_x^2. \quad (7.3)$$

□

Remark 7.1. *Restricting the secular system (7.1), (7.2) to $\mathcal{D} \neq 0$ (because $\sin \alpha \cos \alpha \neq 0$ according to Assertions 7.2 and 7.3) makes it possible to rewrite (7.1) in the form*

$$\xi(A^2 - B^2)\mathcal{D} - \mathcal{F}[\xi^2 + (\Lambda + m_1(A^2 + B^2))^2]/(m_3 - m_1) + A^2 B^2 \mathcal{D}^2 = 0, \quad (7.4)$$

which is the π-periodic function of α. The subsystem (7.2) is also the π-periodic functions of α. However, the original secular system (5.31a) does not possess this feature. This means that (7.2), (7.4) is not equivalent to (5.31a), namely, solving (7.2), (7.4) does not compute correctly the phase-lag difference α, but outputs $\alpha = \alpha_0 + \pi i$, $i \in \mathbb{Z}$. To return the original α, we note that the left-hand side of (5.31a) is the π-periodic function of α and, therefore, substituting $\alpha = \alpha_0 + \pi i$, $i \in \mathbb{N}$ in this left-hand side gives the original values of $\epsilon_x \cos \psi$, $\epsilon_x \sin \psi$ and $\epsilon_y \sin \varphi$, and $\epsilon_y \cos \varphi$. This makes it possible to find ψ and φ and, of course, $\alpha = \varphi - \psi$.

Remark 7.2. *Section 3.1.2.2 introduces counterclockwise and clockwise swirling wave modes whose direction depends on the sign of $\Xi = ab - \bar{a}\bar{b} = AB \sin \alpha$ (see the alternative (5.33)). This means that using the computed α makes it possible to identify counterclockwise ($\sin \alpha > 0$) and clockwise ($\sin \alpha < 0$) swirling waves. According to Assertion 7.1, the analysis, without loss of generality, may assume the counterclockwise orbital forcing with $\delta_1 > 0$. This means that the sign of $\sin \alpha$ should be used to distinguish co- and* counter-directed swirling, that is, $\sin\alpha > 0$ implies the co-directed swirling but $\sin\alpha < 0$ means that swirling occurs in the opposite direction to the forcing orbit.

As for the longitudinal forcing, our numerical experiments show that $C > 0$ for $(m_3 - m_1) > 0$. A numerical-analytical scheme is used to find solutions of (7.2), (7.4). First, the scheme considers $\mathcal{F}$ and $\mathcal{D}$ as functions of $0 < \beta < 1$,

$$\mathcal{F}(\beta) = (m_3 - m_1)\beta, \quad \mathcal{D}(\beta) = (m_3 - m_1)\sqrt{\beta(1-\beta)}, \quad C > 0. \tag{7.5}$$

A simple analysis shows that

$$0 < A < \frac{\epsilon_x}{\xi}, \quad 0 < B^2 \leq \min\left[\frac{1}{\mathcal{D}(\beta)}\left(\frac{\epsilon_x}{A} - \xi\right), \frac{\delta^2 \epsilon_x^2}{(\mathcal{D}(\beta)A^2 - \xi)^2}\right], \tag{7.6}$$

which determines the fixed intervals for A and B but B^2 belongs to the curvilinear trapezoid by (7.6). The first equation (7.2) outputs the two real values $\Lambda_\pm$ for any $0 < \beta < 1$ and A, B^2, satisfying (7.6), as follows

$$\Lambda_\pm = -m_1 A^2 - (m_3 - \mathcal{F}(\beta))B^2 \pm \sqrt{\frac{\epsilon_x^2}{A^2} - (\mathcal{D}(\beta)B^2 + \xi)^2}. \tag{7.7}$$

To solve (7.4), (7.2) for *any fixed A* belonging to the interval (7.6):

1) We introduce a mesh $0 < \beta_1 < \beta_2 < ... < \beta_k < ... < \beta_K < 1$.
2) For any fixed $\beta_k \in \{\beta_n\}$, we solve two equations (that follow from the second equation in (7.2))

$$[\Lambda_\pm + m_1 B^2 + (m_3 - \mathcal{F}(\beta))A^2]^2 + [\mathcal{D}(\beta)A^2 - \xi]^2 = \frac{\delta^2 \epsilon_x^2}{B^2}, \; j = 1, 2, \tag{7.8}$$

(associated with $+$ and $-$ in the expression (7.7), respectively) with respect to B^2 on the interval (7.6); the result will be a set of positive roots $B_{k,j,i}^2 = B_i^2(A, \beta_k, j)$, $j = 1, 2$, for any A and β_k.
3) Every root $B_i(A, \beta_k, j)$ is then substitutes into (7.4):

$$\xi(A^2 - B_{i,k,j}^2)\mathcal{D}(\beta_k) - \mathcal{F}(\beta_k)[\xi^2 + (\Lambda_\pm + m_1(A^2 + B_{i,k,j}^2))^2]/(m_3 - m_1)$$
$$+ A^2 B_{i,k,j}^2 \mathcal{D}^2(\beta_k) = 0 \tag{7.9}$$

to determine the grid (β_k, β_{k+1}), where the left side of (7.9) changes the sign.
4) An iterative procedure for calculating $\beta \in (\beta_k, \beta_{k+1})$ and corresponding $B_i(A, \beta, j)$ is used.

The algorithm calculates numerical solutions for any fixed A. By changing A in (7.6), we obtain the corresponding response curves in $(\sigma/\sigma_{11}, A, B)$-space. For each point on the response curve, one can specify stability and type of the swirling wave mode (co-directed or counter-directed) as commented in Remark 7.2.

7.2 WAVE-AMPLITUDE RESPONSE CURVES FOR ELLIPTIC ORBITS

Applying the numerical method from the previous section (for $\xi > 0$), as well as using analytic solutions from [52] for the case $\xi = 0$, we construct illustrative response curves in the $(\sigma/\sigma_{11}, A, B)$-space for different values of $0 < \delta_1 < 1$.

7.2.1 UNDAMPED SLOSHING

The response curves in the $(\sigma/\sigma_{11}, A, B)$-space for the elliptic tank forcing and the undamped sloshing can be drawn by using the corresponding analytical solution of the secular system (5.12) from [52]; $\xi = \bar{\epsilon}_x = \bar{\epsilon}_y = 0$ and $\epsilon_y = \delta_1 \epsilon_x$, $0 < \delta_1 < 1$. As we remarked in section 5.3.1, the aforementioned paper proved that $\bar{a} = \bar{b}$ for that case ($\Leftrightarrow \sin \psi = \cos \varphi = 0$), but $a \neq b$. The secular system (5.12) can then be rewritten in the form

$$\begin{cases} b\left[(m_1 - m_3)b^2 + \left(\dfrac{\epsilon_x}{a} - (m_1 - m_3)a^2\right)\right] = \delta_1 \epsilon_x, \\ \Lambda = \dfrac{\epsilon_x}{a} - m_1 a^2 - m_3 b^2. \end{cases} \quad (7.10)$$

The first equation of (7.10) is a depressed cubic with respect to b whose coefficient at the linear form is a function of a. The second equality computes the forcing frequency parameter Λ as a function of a and b. Varying a in an admissible range, solving the depressed cubic (finding $b = b(a)$) and computing the frequency parameter from the second equation determines the response curves $(\sigma/\sigma_{11}, A, B) = (\sigma/\sigma_{11}, |a|, |b|)$. When solving the depressed cubic, one should check for the discriminant

$$\triangle_c = -4\left(\dfrac{\epsilon_x}{a(m_1 - m_3)} - a^2\right)^3 - 27\left(\dfrac{\delta_1 \epsilon_x}{m_1 - m_3}\right)^2.$$

Cartano's theorem states that, if $\triangle_c > 0$, then there are three distinct real roots for b, if $\triangle_c = 0$, then the equation has at least one multiple root and all its roots are real, and if $\triangle_c < 0$, then the equation has one real root and two non-real complex conjugate roots. When considering $\triangle_c$ as a function of a, a simple analysis shows that, when $m_3 - m_1 > 0$, there exists only a negative real root $a_* < 0$ of $\triangle_c(a_* = 0)$ so that $\triangle_c(a) > 0$ for $a < a_*$ and $0 < a$ (three real solutions) but $\triangle_c(a) < 0$ for $a_* < a < 0$ (a unique real solution).

When $\delta_1 = 0$ (longitudinal forcing), the depressed cubic in (7.10) has the zero solution ($b = 0$, planar standing waves) and two non-zero solutions with opposite signs ($\pm b \neq 0$, swirling), the sign is responsible for swirling direction (($+$) implies counterclockwise, but ($-$) means clockwise swirling wave modes). Our computations assume $0 < \delta_1 < 1$ (counterclockwise forcing). When $\Xi = ab > 0$, swirling is co-directed with the forcing, but $\Xi = ab < 0$ implies the counter-directed one. The results are demonstrated for $\delta_1 = 0.05, 0.2, 0.3$, $0.45, 0.5$, and 0.95 in Figure 7.1. All points on these response curves correspond to swirling, but some branches are close to the $(\sigma/\sigma_{11}, A)$-plane, which means that the corresponding swirling wave implies a nearly-planar standing

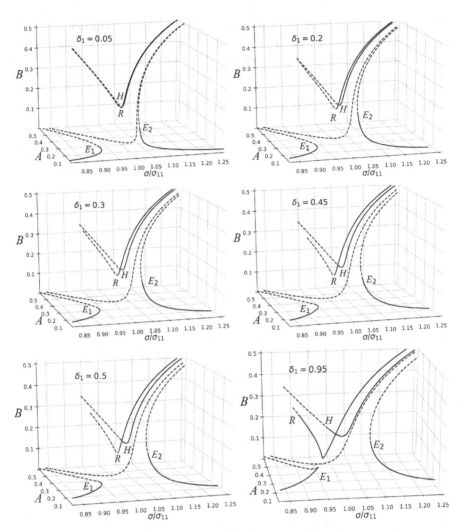

Figure 7.1 The wave-amplitude response curves in $(\sigma/\sigma_{11}, A, B)$-space for the undamped steady-state resonant sloshing due to the elliptic periodic horizontal tank forcing with $\eta_{1a} = 0.01$, $\eta_{2a} = \delta_1 \eta_{1a}$, where $\delta_1 = 0.05, 0.2, 0.3, 0.45, 0.5$, and 0.95. The solid lines correspond to the stable sloshing, but the dashed lines – the instability. The branch containing R implies swirling propagating against the orbital forcing.

mode. This (almost planar/standing) wave is always unstable in the vicinity of the fundamental resonance $\sigma/\sigma_{11} = 1$, where swirling (in the right of $H(R)$) and irregular waves (between points E_1 and $H(R)$) are expected. In contrast to the undamped steady-state waves for the longitudinal forcing (the response curves in Figure 6.1(a)), the branch, which contains H, splits into the two nonconnected branches in Figure 7.1 (see the panels with either $\delta_1 = 0.05$ or $\delta_1 = 0.2$). The new 'split' branch is marked by the point R; points on this branch correspond to the counter-directed swirling. With increasing δ_1, the branches in Figure 7.1 with points H and R continue diverging from each other. And it can be seen that the R-branch 'goes' down to the E_1-branch and even approaches it (see the case with $\delta_1 = 0.95$ in Figure 7.1).

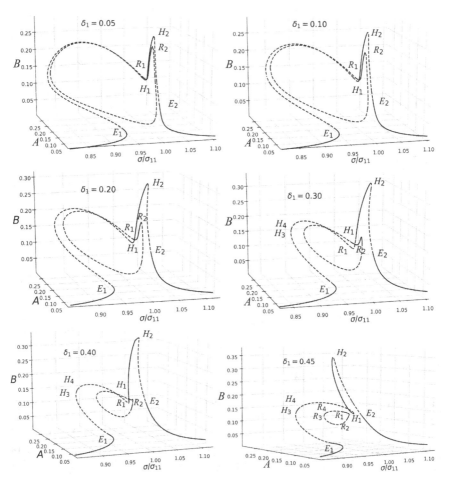

Figure 7.2 The wave-amplitude response curves in the $(\sigma/\sigma_{11}, A, B)$-space for the steady-state resonant damped sloshing. The geometric results should be similar to those in Figure 7.1, but the non-zero damping ($\xi = 0.02$) topologically changes them. The branch with points E_1, H_1, H_2, and E_2 corresponds to the co-directed swirling, but the loop-type branch, which contains $R_1 R_2$ (and a small interval $R_3 R_4$ for $\delta_1 = 0.45$), implies the counter-directed swirling. For these semi-axes ratios, there exists a frequency range where irregular (chaotic, modulated, etc.) waves occur.

The main conclusion from the illustrative response curves in Figure 7.1 is that both co- and counter-directed (to the orbit direction) stable swirling waves exist in the vicinity of the primary resonance $\sigma/\sigma_{11} \approx 1$ as $0 < \delta_1 < 1$. The undamped theory also states that zone of irregular waves disappears with increasing δ_1.

7.2.2 DAMPED SLOSHING

Figure 7.2 illustrates the wave-amplitude response curves of the damped ($\xi = 0.02$ in these computations) resonant steady-state sloshing due to the elliptic tank forcing. Passage from reciprocating to elliptic orbits shows topological changes, which are qualitatively similar to what we observed for the undamped case in Figure 7.1. There is a splitting of the arc-type subbranch $E_1 H_1 H_2 E_2$, which is depicted in Figure 6.1(b); we see that in Figure 7.2 for $\delta_1 = 0.05$. In addition to the arc-type subbranch $E_1 H_1 H_2 E_2$, the loop-type branch $R_1 R_2$ appears. The latter does not intersect $E_1 H_1 H_2 E_2$ and is responsible for the counter-directed swirling. With an increase of δ_1 (about $\delta_1 = 0.3$) islands of stability ($H_3 H_4$ on the branch $E_1 H_1 H_2 E_2$) may appear. The loop-type branch $R_1 R_2$ decreases. With a further increase of δ_1, in the vicinity of $\delta_1 = 0.45$, an additional stability range $R_3 R_4$ on the branch with R_1 and R_2 may also happen. The loop-type branch $R_1 R_2 (R_1 R_2 R_3 R_4)$ diminishes and, furthermore, disappears for a certain δ_1 between 0.45 and 0.5. As a result, the loop-type branch does not exist for $\delta_1 = 0.5$ in Figure 7.3. With increasing to $\delta_1 = 0.8$, the range $H_3 H_4$ increases, connecting, eventually, with $H_1 H_2$ and point E_1.

Figure 7.4 illustrates the wave-amplitude response curves for the almost rotary and rotary tank forcing. It can be seen that transition to the circular forcing not only makes the counter-directed swirling impossible but also makes impossible the frequency range where irregular waves occur. The corresponding frequency range vanishes at approximately $\delta_1 = 0.56$.

In *summary*, the linear viscous damping significantly affects the steady-state resonant sloshing in the orbitally excited tank with elliptic tank trajectories. The most interesting fact is that the damping, even being relatively small, prevents the counter-directed swirling as $\delta_1 \to 1$. This fact is supported by experimental observations in [200]. Another important fact is that irregular waves become impossible with increasing the semi-axes ratio.

7.3 PHASE LAGS FOR THE DAMPED SLOSHING

The non-zero damping qualitatively changes the phase-lag response curves. The undamped steady-state resonant sloshing due to the orbital forcing is characterised by piecewise ψ and φ, so that, when $0 < \delta_1 < 1$, the equality holds true $\sin\psi = \cos\varphi = 0$ in (5.30b), which is a consequence of $\bar{a} = \bar{b} = 0$ proven in [52]. However, the phase lags become complex functions of the forcing frequency σ/σ_{11} for the non-zero damping $\xi > 0$.

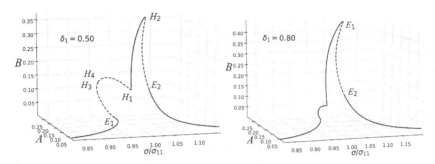

Figure 7.3 The same as in Figure 7.2 but for $\delta_1 = 0.5$ and 0.8. Specifically, the semi-axes ratio $\delta_1 = 0.5$ eliminates the counter-directed swirling, but irregular waves are possible in the frequency range between E_1 and H_1.

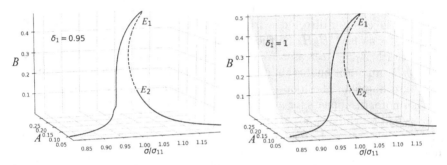

Figure 7.4 The same as in Figure 7.2 but for $\delta_1 = 0.95$ and 1. For the nearly-circular forcing, only co-directed stable swirling is expected.

Figure 7.5 illustrates the phase-lag ψ/π response curves for elliptic orbital tank excitations and different semi-axes ratios. Comparing these phase-lag response curves with analogous results for the longitudinal forcing in Figure 6.2 shows that the connected branch with $\delta_1 = 0$ splits (see the case $\delta_1 = 0.05$ in the first panel) into a continuous curve $E_1 H_1 H_2 E_2$ and the loop-type curve $R_1 R_2$. Points on the first curve imply the co-directed swirling but the loop-type branch is responsible for the counter-directed swirling. As earlier, we see how the counter-directed swirling disappears with increasing δ_1 (it is absent in the panel with $\delta_1 = 0.5$), and a further increase of the semi-axes ratio eliminates irregular waves. The latter waves still exist in the frequency interval determined by E_1 and H_1 with $\delta_1 = 0.5$ but we do not see the corresponding frequency range for $\delta_1 = 0.95$. Specifically, the phase-lag response curves in Figure 7.5 also demonstrate an emergence of additional 'islands' of stability on $E_1 H_1 H_2 E_2$ and on the loop-type branch $R_1 R_2$.

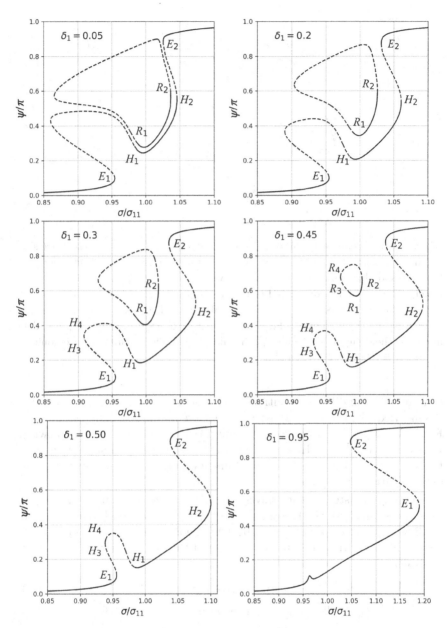

Figure 7.5 The phase-lag ψ/π for different values of the semi-axes ratio δ_1. The considered cases correspond to the wave-amplitude response curves in Figure 7.1 and adopt the same notations.

7.4 ROTARY TANK MOTIONS

Let us study what happens when $\delta_1 = 1$ (the rotary orbital tank forcing). As explained in section 5.3.1 for the undamped case, the circular forcing causes a

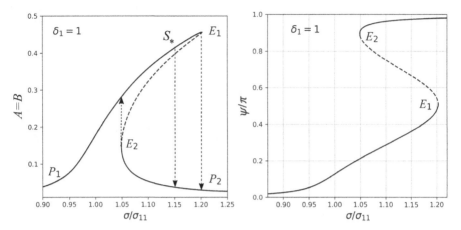

Figure 7.6 The wave-amplitude response curves of the steady-state resonance sloshing in the $(\sigma/\sigma_{11}, A = B)$-plane for the rotary orbital tank forcing (left) and the corresponding phase-lag response curves in the $(\sigma/\sigma_{11}, \psi/\pi)$-plane (the right panel). The solid lines correspond to the stable sloshing, but the dashed lines imply the instability. Between E_1 and E_2, we observe a hysteresis where two steady-state wave motions co-exist. The coordinates of bifurcation point E_1 are determined by the damping coefficient ξ (formula (7.13)). The calculations are done with $\eta_{1a} = \eta_{2a} = 0.0066$, $h = 1.5$ and, $\xi = 0.0128$.

continuum set of different solutions of the secular system (5.12) for each fixed forcing frequency σ/σ_{11}. Moreover, the wave-amplitude response curves in Figure 7.1 show that both co- and counter-directed stable undamped swirling waves exist as $\delta_1 \to 1$. However, according to the damped sloshing expectations in Figures 7.2–7.4 and in experiments, only co-directed swirling waves exist in this limiting case.

To study the steady-state waves with the non-zero damping by using the secular system (7.4), (7.2) with $\delta_1 = 1$ and $\xi > 0$, let us recall that $C \neq 0$, but the limit $C = \tan\alpha \to \pm\infty$ is possible (Assertion 7.3 assumes $0 < \delta_1 < 1$). The last limit transforms (7.4) to $A^2 = B^2$, and the two equations (7.2) become equivalent

$$A = B > 0, \quad A^2(\Lambda + (m_1 + m_3)A^2)^2 + \xi^2) = \epsilon_x^2 = \epsilon_y^2 = \epsilon^2, \quad (7.11)$$

$\mathcal{D} = \mathcal{F} = 0$, which allows us to recover ψ and $\varphi - \psi = \pi/2$ from

$$A(\Lambda + (m_1 + m_3)A^2) = \epsilon\cos\psi, \quad A\xi = \epsilon\sin\psi. \quad (7.12)$$

According to (6.10), m_1 and m_3 satisfy the inequalities $m_1 < 0$, $m_3 > 0$ for $h \gtrsim 1.05$, therefore, the corresponding wave-amplitude response curves that follow from (7.11) should demonstrate the damped hard-spring behaviour, which is illustrated in Figure 7.6. The non-zero damping coefficient $\xi > 0$ affects the position of E_1. The nondimensional forcing frequency

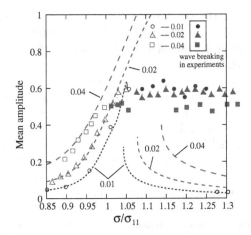

Figure 7.7 The theoretical (lines) and experimental (symbols) nondimensional amplitude parameters of the steady-state waves $(\zeta_{max} - \zeta_{min})/2$ for the rotary tank forcing. Experiments [200, 201] were carried out with a gradual increase of σ, that is, by moving along $P_1 E$ in Figure 7.6. The theoretical curves are indicated by values of $\eta_{1a} = \eta_{2a} = 0.01, 0.02$, and 0.04. The calculations are done with the damping coefficient $\xi = 0.034$ (main text explains this value). The empty symbols correspond to experimental observations, where a strong wave breaking and overturning occur, the Prandtl mass-transport phenomenon [71, 96, 192] is also expected to affect the free-surface elevations.

$(\sigma/\sigma_{11})_\star$ at the bifurcation point $S_\star$ (located between E_1 and E_2) can be found from from the inequality

$$(\sigma/\sigma_{11})_\star \leq \left(1 - (m_1 + m_3)(\epsilon/\xi)^2\right)^{-1/2}, \quad \epsilon = \epsilon_x = \epsilon_y \qquad (7.13)$$

whose right-hand side determines the upper bound of the hysteresis zone, i.e., the abscissa of E_1 in the left panel of Figure 7.6. One can see that this upper bound increases with increasing ϵ/ξ. Practically, this means that one can use experimental estimates of E_1 to get a rough estimate of the damping coefficient from above. Formulas (3.74)–(3.79) provide the lower bound of the damping coefficient.

For the experimental cases in [200, 201], the formulas (3.74)–(3.79) compute the lower bound $0.031 = 2\xi_{11}^{(0)} \leq \xi$. Figure 7.7 shows the measured mean wave amplitude from these experiments for the forcing amplitudes $\eta_{1a} = \eta_{2a} = 0.01, 0.02$, and 0.04. The filled symbols are used to indicate breaking waves and overturning, which accompany the corresponding model test. These free-surface phenomena were especially strong for the forcing amplitude 0.04, thus, only the model tests with $\eta_{1a} = \eta_{2a} = 0.01$ and 0.02 can be employed to get a rough estimate of the upper bound (position of E_1). According to Figure 7.7, the jump from upper to lower branch for the forcing amplitude 0.01 occurs at $\sigma/\sigma_{11} = 1.27$, but the same jump for $\eta_{1a} = \eta_{2a} = 0.02$ was detected at $\sigma/\sigma_{11} = 1.45$ (it is not presented in Figure 7.7, but [200,201] report this value). These two experimental abscissas of E_1 compute (from (7.13)) the upper bound as $\xi \leq 0.034$. This means that, in the *points, where the Narimanov-Moiseev approximation is supported by measurements* [200, 201], the viscous damping should be relatively small $0.031 \leq \xi \leq 0.034$, and it is, mainly, associated with a viscous laminar surface layer on wetted walls of the tank.

Usage of the experimental value $\xi = 0.034$ in our theoretical steady-state analysis makes it possible to compare theoretical mean wave elevations with the measured values from [200,201]. Figure 7.7 demonstrates a good agreement between the Narimanov-Moiseev steady-state theory and experiments in the frequency ranges where no overturning and breaking waves were observed. Unfortunately, the experimental observations report breaking wave and other free-surface phenomena (solid symbols). Just for these cases, the theory is not supported by the measurements. The breaking wave effect may be modelled by a speculative increase of the total damping in the hydrodynamic system, but this increase is not related to the lowest damping ratio ξ_{11}; the damping is then due to energy content to the higher natural sloshing modes. The mechanism is rather complicated. It was in details described in [59]. The description requires a special type of the adaptive modal system.

8 Tables of hydrodynamic coefficients

The linear modal theory in section 3.2.2 is characterised by the hydrodynamic coefficients in the modal equations (3.56), (3.60), (3.61), and (3.64), which are easily computable when we know the roots of $J_1'(k_{1n}) = 0$. The roots are presented in the second column of Table 8.1. The nonlinear modal system of Narimanov-Moiseev–type also requires the roots $J_M'(k_{Mi}) = 0$ (for $M = 0, 1, 2, 3$), $J_M(k_{Mi})$, and the hydrodynamic coefficients at nonlinear terms. These are presented in other columns of Table 8.1 and in Tables 8.2–8.17.

Table 8.1
Roots of the equation $J_M'(k_{Mn}) = 0$, $n = 1, \ldots, 10$, $M = 0, \ldots, 3$.

i	k_{0i}	k_{1i}	k_{2i}	k_{3i}
1	3.83170597021	1.84118378134	3.05423692823	4.20118894121
2	7.01558666982	5.33144277353	6.70613319416	8.01523659838
3	10.1734681351	8.53631636635	9.96946782309	11.3459243107
4	13.3236919363	11.7060049026	13.1703708560	14.5858482862
5	16.4706300509	14.8635886339	16.3475223183	17.7887478661
6	19.6158585105	18.0155278627	19.5129127825	20.9724769365
7	22.7600843806	21.1643698592	22.6715817725	24.1448974329
8	25.9036720876	24.3113268572	25.8260371418	27.3100579302
9	29.0468285349	27.4570505711	28.9776727730	30.4702688063
10	32.1896799110	30.6019229727	32.1273270204	33.6269491828

Table 8.2
Values of $J_M(k_{Mn}) = 0$, $n = 1, \ldots, 4$, $M = 0, \ldots, 3$.

i	$J_0(k_{0i})$	$J_1(k_{1i})$	$J_2(k_{2i})$	$J_3(k_{3i})$
1	−0.402759395703	0.581865224282	0.486498682269	0.434394426841
2	0.300115752526	−0.346126201854	−0.313530445158	−0.291161281463
3	−0.249704877058	0.273299941633	0.254744158212	0.240738174862
4	0.218359407248	−0.233304417171	−0.220881582151	−0.210965203388

Table 8.3

h	d_1	d_2	m_1	m_3	$d_3^{(1)}$	$d_3^{(2)}$	$d_3^{(3)}$	$d_3^{(4)}$	$d_4^{(1)}$	$d_4^{(2)}$	$d_4^{(3)}$	$d_4^{(4)}$
1.05	1.72752	−1.10438	−1.41649	2.93796	1.90086	0.07933	−0.02760	0.01455	0.17848	−0.01001	0.00317	−0.00143
1.10	1.69921	−1.12301	−1.61470	3.12004	1.90652	0.07893	−0.02743	0.01446	0.18903	−0.00993	0.00316	−0.00143
1.15	1.67605	−1.13820	−1.68294	3.17862	1.91131	0.07860	−0.02729	0.01438	0.19776	−0.00986	0.00315	−0.00143
1.20	1.65707	−1.15061	−1.72090	3.21058	1.91534	0.07833	−0.02717	0.01431	0.20498	−0.00981	0.00314	−0.00142
1.25	1.64148	−1.16079	−1.74600	3.23187	1.91872	0.07810	−0.02708	0.01426	0.21096	−0.00976	0.00314	−0.00142
1.30	1.62865	−1.16915	−1.76403	3.24745	1.92156	0.07792	−0.02700	0.01422	0.21591	−0.00973	0.00313	−0.00142
1.35	1.61808	−1.17602	−1.77760	3.25942	1.92394	0.07776	−0.02693	0.01418	0.22002	−0.00969	0.00312	−0.00142
1.40	1.60936	−1.18168	−1.78810	3.26888	1.92593	0.07763	−0.02688	0.01415	0.22342	−0.00967	0.00312	−0.00142
1.45	1.60215	−1.18634	−1.79638	3.27646	1.92760	0.07753	−0.02683	0.01412	0.22624	−0.00964	0.00312	−0.00142
1.50	1.59619	−1.19020	−1.80299	3.28262	1.92899	0.07744	−0.02680	0.01410	0.22858	−0.00963	0.00311	−0.00141
1.55	1.59126	−1.19338	−1.80832	3.28764	1.93015	0.07737	−0.02676	0.01408	0.23052	−0.00961	0.00311	−0.00141
1.60	1.58717	−1.19601	−1.81264	3.29176	1.93112	0.07731	−0.02674	0.01407	0.23213	−0.00960	0.00311	−0.00141
1.65	1.58379	−1.19819	−1.81617	3.29515	1.93192	0.07726	−0.02672	0.01406	0.23346	−0.00959	0.00311	−0.00141
1.70	1.58098	−1.20000	−1.81906	3.29795	1.93259	0.07721	−0.02670	0.01405	0.23457	−0.00958	0.00311	−0.00141
1.75	1.57866	−1.20149	−1.82143	3.30026	1.93315	0.07718	−0.02668	0.01404	0.23549	−0.00957	0.00311	−0.00141
1.80	1.57673	−1.20273	−1.82338	3.30217	1.93362	0.07715	−0.02667	0.01403	0.23626	−0.00956	0.00310	−0.00141
1.85	1.57512	−1.20376	−1.82499	3.30376	1.93401	0.07713	−0.02666	0.01403	0.23689	−0.00956	0.00310	−0.00141
1.90	1.57379	−1.20462	−1.82632	3.30507	1.93433	0.07711	−0.02665	0.01402	0.23742	−0.00955	0.00310	−0.00141
1.95	1.57269	−1.20532	−1.82742	3.30616	1.93460	0.07709	−0.02665	0.01402	0.23786	−0.00955	0.00310	−0.00141
2.00	1.57177	−1.20591	−1.82833	3.30706	1.93483	0.07708	−0.02664	0.01402	0.23822	−0.00955	0.00310	−0.00141
2.05	1.57101	−1.20640	−1.82909	3.30781	1.93501	0.07706	−0.02664	0.01401	0.23852	−0.00954	0.00310	−0.00141
2.10	1.57038	−1.20680	−1.82972	3.30844	1.93517	0.07705	−0.02663	0.01401	0.23877	−0.00954	0.00310	−0.00141
2.15	1.56985	−1.20714	−1.83024	3.30895	1.93530	0.07705	−0.02663	0.01401	0.23898	−0.00954	0.00310	−0.00141
2.20	1.56941	−1.20742	−1.83067	3.30938	1.93541	0.07704	−0.02663	0.01401	0.23916	−0.00954	0.00310	−0.00141
2.25	1.56905	−1.20765	−1.83103	3.30974	1.93549	0.07703	−0.02662	0.01400	0.23930	−0.00954	0.00310	−0.00141
2.30	1.56875	−1.20785	−1.83132	3.31004	1.93557	0.07703	−0.02662	0.01400	0.23942	−0.00954	0.00310	−0.00141
2.40	1.56829	−1.20814	−1.83178	3.31049	1.93568	0.07702	−0.02662	0.01400	0.23960	−0.00954	0.00310	−0.00141
2.50	1.56797	−1.20834	−1.83209	3.31080	1.93576	0.07702	−0.02662	0.01400	0.23973	−0.00954	0.00310	−0.00141
2.60	1.56775	−1.20848	−1.83231	3.31102	1.93582	0.07702	−0.02661	0.01400	0.23982	−0.00953	0.00310	−0.00141
2.70	1.56760	−1.20858	−1.83245	3.31117	1.93585	0.07701	−0.02661	0.01400	0.23988	−0.00953	0.00310	−0.00141
2.80	1.56749	−1.20865	−1.83256	3.31127	1.93588	0.07701	−0.02661	0.01400	0.23992	−0.00953	0.00310	−0.00141
2.90	1.56742	−1.20869	−1.83263	3.31134	1.93590	0.07701	−0.02661	0.01400	0.23995	−0.00953	0.00310	−0.00141
3.00	1.56737	−1.20873	−1.83268	3.31139	1.93591	0.07701	−0.02661	0.01400	0.23997	−0.00953	0.00310	−0.00141
3.10	1.56733	−1.20875	−1.83271	3.31142	1.93592	0.07701	−0.02661	0.01400	0.23998	−0.00953	0.00310	−0.00141

Table 8.4

h	$d_5^{(1)}$	$d_5^{(2)}$	$d_5^{(3)}$	$d_5^{(4)}$	$d_6^{(1)}$	$d_6^{(2)}$	$d_6^{(3)}$	$d_6^{(4)}$	$d_7^{(1)}$	$d_7^{(2)}$	$d_7^{(3)}$	$d_7^{(4)}$
1.05	−2.32107	0.12606	−0.05040	0.02769	0.08778	−0.01593	0.00574	−0.00271	−1.33092	−0.18866	0.09583	−0.06495
1.10	−2.31920	0.12541	−0.05009	0.02751	0.08031	−0.01582	0.00572	−0.00270	−1.30950	−0.18629	0.09462	−0.06412
1.15	−2.31774	0.12486	−0.04983	0.02736	0.07410	−0.01572	0.00570	−0.00269	−1.29189	−0.18433	0.09362	−0.06343
1.20	−2.31658	0.12442	−0.04962	0.02724	0.06893	−0.01564	0.00569	−0.00269	−1.27737	−0.18272	0.09279	−0.06287
1.25	−2.31565	0.12405	−0.04945	0.02713	0.06463	−0.01558	0.00567	−0.00268	−1.26539	−0.18140	0.09211	−0.06240
1.30	−2.31491	0.12374	−0.04930	0.02705	0.06105	−0.01552	0.00566	−0.00268	−1.25550	−0.18030	0.09155	−0.06202
1.35	−2.31432	0.12349	−0.04918	0.02698	0.05807	−0.01548	0.00565	−0.00268	−1.24732	−0.17939	0.09109	−0.06170
1.40	−2.31384	0.12328	−0.04908	0.02692	0.05559	−0.01544	0.00565	−0.00268	−1.24055	−0.17864	0.09071	−0.06144
1.45	−2.31345	0.12310	−0.04900	0.02687	0.05352	−0.01541	0.00564	−0.00267	−1.23494	−0.17802	0.09039	−0.06122
1.50	−2.31313	0.12296	−0.04893	0.02683	0.05180	−0.01538	0.00564	−0.00267	−1.23030	−0.17750	0.09012	−0.06104
1.55	−2.31287	0.12284	−0.04887	0.02680	0.05037	−0.01535	0.00563	−0.00267	−1.22644	−0.17707	0.08990	−0.06089
1.60	−2.31266	0.12274	−0.04882	0.02677	0.04918	−0.01534	0.00563	−0.00267	−1.22325	−0.17671	0.08972	−0.06077
1.65	−2.31248	0.12266	−0.04879	0.02675	0.04819	−0.01532	0.00563	−0.00267	−1.22059	−0.17642	0.08957	−0.06066
1.70	−2.31234	0.12259	−0.04875	0.02673	0.04737	−0.01531	0.00562	−0.00267	−1.21839	−0.17617	0.08945	−0.06058
1.75	−2.31222	0.12253	−0.04873	0.02671	0.04668	−0.01530	0.00562	−0.00267	−1.21656	−0.17597	0.08934	−0.06051
1.80	−2.31212	0.12248	−0.04870	0.02670	0.04611	−0.01529	0.00562	−0.00267	−1.21504	−0.17580	0.08926	−0.06045
1.85	−2.31204	0.12244	−0.04868	0.02669	0.04564	−0.01528	0.00562	−0.00267	−1.21378	−0.17566	0.08918	−0.06040
1.90	−2.31197	0.12241	−0.04867	0.02668	0.04524	−0.01527	0.00562	−0.00267	−1.21273	−0.17554	0.08912	−0.06036
1.95	−2.31192	0.12238	−0.04866	0.02667	0.04492	−0.01527	0.00562	−0.00266	−1.21186	−0.17545	0.08908	−0.06032
2.00	−2.31187	0.12236	−0.04864	0.02667	0.04464	−0.01526	0.00561	−0.00266	−1.21114	−0.17536	0.08903	−0.06029
2.05	−2.31183	0.12234	−0.04864	0.02666	0.04442	−0.01526	0.00561	−0.00266	−1.21054	−0.17530	0.08900	−0.06027
2.10	−2.31180	0.12233	−0.04863	0.02666	0.04423	−0.01526	0.00561	−0.00266	−1.21004	−0.17524	0.08897	−0.06025
2.15	−2.31178	0.12231	−0.04862	0.02665	0.04407	−0.01526	0.00561	−0.00266	−1.20962	−0.17519	0.08895	−0.06023
2.20	−2.31175	0.12230	−0.04862	0.02665	0.04394	−0.01525	0.00561	−0.00266	−1.20928	−0.17516	0.08893	−0.06022
2.25	−2.31174	0.12229	−0.04861	0.02665	0.04383	−0.01525	0.00561	−0.00266	−1.20899	−0.17512	0.08891	−0.06021
2.30	−2.31172	0.12228	−0.04861	0.02665	0.04374	−0.01525	0.00561	−0.00266	−1.20875	−0.17510	0.08890	−0.06020
2.40	−2.31170	0.12227	−0.04860	0.02664	0.04360	−0.01525	0.00561	−0.00266	−1.20838	−0.17506	0.08888	−0.06019
2.50	−2.31168	0.12227	−0.04860	0.02664	0.04351	−0.01525	0.00561	−0.00266	−1.20813	−0.17503	0.08886	−0.06018
2.60	−2.31167	0.12226	−0.04860	0.02664	0.04344	−0.01524	0.00561	−0.00266	−1.20796	−0.17501	0.08885	−0.06017
2.70	−2.31166	0.12226	−0.04860	0.02664	0.04339	−0.01524	0.00561	−0.00266	−1.20784	−0.17499	0.08885	−0.06016
2.80	−2.31166	0.12225	−0.04859	0.02664	0.04336	−0.01524	0.00561	−0.00266	−1.20775	−0.17499	0.08884	−0.06016
2.90	−2.31166	0.12225	−0.04859	0.02664	0.04334	−0.01524	0.00561	−0.00266	−1.20770	−0.17498	0.08884	−0.06016
3.00	−2.31165	0.12225	−0.04859	0.02664	0.04333	−0.01524	0.00561	−0.00266	−1.20766	−0.17497	0.08883	−0.06016
3.10	−2.31165	0.12225	−0.04859	0.02664	0.04332	−0.01524	0.00561	−0.00266	−1.20763	−0.17497	0.08883	−0.06016

Table 8.5

h	$d_8^{(1)}$	$d_8^{(2)}$	$d_8^{(3)}$	$d_8^{(4)}$	$d_9^{(1)}$	$d_9^{(2)}$	$d_9^{(3)}$	$d_{10}^{(4)}$	$d_{10}^{(1)}$	$d_{10}^{(2)}$	$d_{10}^{(3)}$	$d_{10}^{(4)}$
1.05	1.35361	−0.15688	0.08914	−0.06244	0.30773	−0.03802	0.01792	−0.01068	0.09518	−0.03165	0.01654	−0.01021
1.10	1.33534	−0.15490	0.08801	−0.06164	0.32390	−0.03745	0.01773	−0.01058	0.08650	−0.03121	0.01636	−0.01011
1.15	1.32029	−0.15328	0.08708	−0.06098	0.33709	−0.03699	0.01756	−0.01050	0.07935	−0.03084	0.01622	−0.01003
1.20	1.30787	−0.15194	0.08631	−0.06044	0.34787	−0.03660	0.01743	−0.01043	0.07346	−0.03053	0.01609	−0.00996
1.25	1.29761	−0.15084	0.08568	−0.05999	0.35669	−0.03628	0.01732	−0.01037	0.06860	−0.03028	0.01599	−0.00991
1.30	1.28912	−0.14993	0.08516	−0.05962	0.36393	−0.03602	0.01723	−0.01032	0.06459	−0.03007	0.01591	−0.00986
1.35	1.28209	−0.14917	0.08473	−0.05931	0.36988	−0.03580	0.01715	−0.01028	0.06126	−0.02989	0.01584	−0.00982
1.40	1.27626	−0.14855	0.08437	−0.05906	0.37478	−0.03562	0.01709	−0.01025	0.05851	−0.02975	0.01578	−0.00979
1.45	1.27143	−0.14803	0.08407	−0.05885	0.37881	−0.03547	0.01704	−0.01022	0.05623	−0.02963	0.01574	−0.00977
1.50	1.26743	−0.14760	0.08383	−0.05868	0.38214	−0.03534	0.01700	−0.01020	0.05433	−0.02953	0.01570	−0.00974
1.55	1.26410	−0.14724	0.08362	−0.05853	0.38488	−0.03524	0.01696	−0.01018	0.05276	−0.02945	0.01566	−0.00973
1.60	1.26134	−0.14695	0.08346	−0.05841	0.38715	−0.03515	0.01693	−0.01016	0.05146	−0.02938	0.01564	−0.00971
1.65	1.25904	−0.14670	0.08332	−0.05831	0.38903	−0.03508	0.01690	−0.01015	0.05038	−0.02932	0.01561	−0.00970
1.70	1.25714	−0.14650	0.08320	−0.05823	0.39058	−0.03502	0.01688	−0.01014	0.04948	−0.02927	0.01559	−0.00969
1.75	1.25556	−0.14633	0.08310	−0.05816	0.39187	−0.03497	0.01687	−0.01013	0.04873	−0.02924	0.01558	−0.00968
1.80	1.25424	−0.14619	0.08302	−0.05810	0.39294	−0.03493	0.01685	−0.01012	0.04811	−0.02920	0.01557	−0.00967
1.85	1.25315	−0.14607	0.08295	−0.05806	0.39382	−0.03489	0.01684	−0.01012	0.04759	−0.02918	0.01556	−0.00967
1.90	1.25224	−0.14597	0.08290	−0.05802	0.39455	−0.03486	0.01683	−0.01011	0.04717	−0.02915	0.01555	−0.00966
1.95	1.25148	−0.14589	0.08285	−0.05799	0.39516	−0.03484	0.01682	−0.01011	0.04681	−0.02913	0.01554	−0.00966
2.00	1.25085	−0.14582	0.08281	−0.05796	0.39567	−0.03482	0.01682	−0.01010	0.04651	−0.02912	0.01554	−0.00965
2.05	1.25033	−0.14577	0.08278	−0.05794	0.39609	−0.03480	0.01681	−0.01010	0.04627	−0.02911	0.01553	−0.00965
2.10	1.24989	−0.14572	0.08276	−0.05792	0.39643	−0.03479	0.01681	−0.01010	0.04606	−0.02909	0.01553	−0.00965
2.15	1.24953	−0.14568	0.08273	−0.05790	0.39672	−0.03478	0.01680	−0.01010	0.04589	−0.02909	0.01552	−0.00965
2.20	1.24923	−0.14565	0.08271	−0.05789	0.39696	−0.03477	0.01680	−0.01009	0.04575	−0.02908	0.01552	−0.00965
2.25	1.24898	−0.14562	0.08270	−0.05788	0.39716	−0.03476	0.01680	−0.01009	0.04563	−0.02907	0.01552	−0.00964
2.30	1.24877	−0.14560	0.08269	−0.05787	0.39733	−0.03475	0.01679	−0.01009	0.04553	−0.02907	0.01551	−0.00964
2.40	1.24846	−0.14557	0.08267	−0.05785	0.39758	−0.03474	0.01679	−0.01009	0.04538	−0.02906	0.01551	−0.00964
2.50	1.24824	−0.14555	0.08265	−0.05784	0.39775	−0.03474	0.01679	−0.01009	0.04528	−0.02905	0.01551	−0.00964
2.60	1.24809	−0.14553	0.08264	−0.05784	0.39787	−0.03473	0.01679	−0.01009	0.04521	−0.02905	0.01550	−0.00964
2.70	1.24798	−0.14552	0.08264	−0.05783	0.39796	−0.03473	0.01678	−0.01009	0.04516	−0.02905	0.01550	−0.00964
2.80	1.24791	−0.14551	0.08263	−0.05783	0.39801	−0.03473	0.01678	−0.01009	0.04512	−0.02904	0.01550	−0.00964
2.90	1.24786	−0.14550	0.08263	−0.05783	0.39805	−0.03472	0.01678	−0.01009	0.04510	−0.02904	0.01550	−0.00964
3.00	1.24782	−0.14550	0.08263	−0.05783	0.39808	−0.03472	0.01678	−0.01009	0.04508	−0.02904	0.01550	−0.00964
3.10	1.24780	−0.14550	0.08263	−0.05783	0.39810	−0.03472	0.01678	−0.01009	0.04507	−0.02904	0.01550	−0.00964

Tables of hydrodynamic coefficients 127

Table 8.6

h	$d_{11,1}$	$d_{11,2}$	$d_{11,3}$	$d_{11,4}$	$d_{12,1}$	$d_{12,2}$	$d_{12,3}$	$d_{12,4}$	$d_{13,1}^{(1)}$	$d_{13,1}^{(2)}$	$d_{13,1}^{(3)}$	$d_{13,1}^{(4)}$
1.05	0.13459	0.02097	−0.01122	−0.03481	1.61640	0.46805	−0.22734	0.11032	0.69629	−1.58647	0.16138	−0.06728
1.10	0.11553	0.01986	−0.01083	−0.03473	1.54454	0.45629	−0.22186	0.10668	0.71496	−1.57966	0.16043	−0.06685
1.15	0.10038	0.01896	−0.01051	−0.03467	1.48686	0.44673	−0.21740	0.10372	0.73034	−1.57402	0.15965	−0.06650
1.20	0.08829	0.01822	−0.01025	−0.03461	1.44035	0.43893	−0.21377	0.10132	0.74302	−1.56934	0.15900	−0.06620
1.25	0.07858	0.01762	−0.01003	−0.03456	1.40271	0.43255	−0.21080	0.09935	0.75350	−1.56547	0.15846	−0.06596
1.30	0.07076	0.01713	−0.00986	−0.03451	1.37213	0.42731	−0.20836	0.09775	0.76217	−1.56226	0.15801	−0.06576
1.35	0.06444	0.01673	−0.00972	−0.03448	1.34722	0.42301	−0.20636	0.09642	0.76935	−1.55959	0.15764	−0.06559
1.40	0.05931	0.01640	−0.00960	−0.03445	1.32688	0.41947	−0.20472	0.09534	0.77530	−1.55737	0.15734	−0.06545
1.45	0.05514	0.01613	−0.00950	−0.03442	1.31022	0.41656	−0.20336	0.09444	0.78023	−1.55554	0.15708	−0.06534
1.50	0.05174	0.01591	−0.00942	−0.03440	1.29656	0.41415	−0.20224	0.09370	0.78432	−1.55401	0.15687	−0.06524
1.55	0.04896	0.01572	−0.00936	−0.03438	1.28534	0.41216	−0.20132	0.09309	0.78772	−1.55274	0.15669	−0.06516
1.60	0.04669	0.01557	−0.00930	−0.03437	1.27611	0.41051	−0.20055	0.09259	0.79054	−1.55169	0.15655	−0.06510
1.65	0.04482	0.01544	−0.00926	−0.03435	1.26850	0.40915	−0.19992	0.09217	0.79288	−1.55081	0.15643	−0.06504
1.70	0.04328	0.01534	−0.00922	−0.03434	1.26222	0.40802	−0.19939	0.09183	0.79483	−1.55008	0.15633	−0.06500
1.75	0.04202	0.01525	−0.00919	−0.03433	1.25704	0.40709	−0.19896	0.09154	0.79645	−1.54948	0.15624	−0.06496
1.80	0.04098	0.01518	−0.00916	−0.03433	1.25275	0.40631	−0.19860	0.09130	0.79779	−1.54897	0.15617	−0.06493
1.85	0.04012	0.01512	−0.00914	−0.03432	1.24921	0.40567	−0.19830	0.09111	0.79891	−1.54855	0.15611	−0.06490
1.90	0.03942	0.01508	−0.00913	−0.03431	1.24628	0.40514	−0.19805	0.09094	0.79984	−1.54820	0.15607	−0.06488
1.95	0.03883	0.01503	−0.00911	−0.03431	1.24385	0.40469	−0.19785	0.09081	0.80062	−1.54792	0.15603	−0.06486
2.00	0.03835	0.01500	−0.00910	−0.03431	1.24183	0.40432	−0.19768	0.09070	0.80126	−1.54767	0.15599	−0.06484
2.05	0.03795	0.01497	−0.00909	−0.03430	1.24016	0.40402	−0.19753	0.09060	0.80179	−1.54747	0.15596	−0.06483
2.10	0.03762	0.01495	−0.00908	−0.03430	1.23878	0.40376	−0.19742	0.09052	0.80224	−1.54731	0.15594	−0.06482
2.15	0.03734	0.01493	−0.00907	−0.03430	1.23763	0.40355	−0.19732	0.09046	0.80261	−1.54717	0.15592	−0.06481
2.20	0.03711	0.01491	−0.00907	−0.03430	1.23668	0.40338	−0.19724	0.09041	0.80291	−1.54705	0.15591	−0.06481
2.25	0.03692	0.01490	−0.00906	−0.03429	1.23589	0.40323	−0.19717	0.09036	0.80317	−1.54696	0.15589	−0.06480
2.30	0.03677	0.01489	−0.00906	−0.03429	1.23523	0.40311	−0.19711	0.09033	0.80338	−1.54688	0.15588	−0.06480
2.40	0.03653	0.01487	−0.00905	−0.03429	1.23423	0.40293	−0.19703	0.09027	0.80371	−1.54676	0.15586	−0.06479
2.50	0.03637	0.01486	−0.00905	−0.03429	1.23354	0.40280	−0.19697	0.09023	0.80393	−1.54667	0.15585	−0.06478
2.60	0.03625	0.01485	−0.00905	−0.03429	1.23307	0.40271	−0.19693	0.09020	0.80409	−1.54662	0.15585	−0.06478
2.70	0.03618	0.01485	−0.00904	−0.03429	1.23274	0.40265	−0.19690	0.09018	0.80419	−1.54657	0.15584	−0.06478
2.80	0.03612	0.01484	−0.00904	−0.03429	1.23251	0.40261	−0.19688	0.09017	0.80427	−1.54655	0.15584	−0.06477
2.90	0.03608	0.01484	−0.00904	−0.03429	1.23235	0.40258	−0.19686	0.09016	0.80432	−1.54653	0.15583	−0.06477
3.00	0.03606	0.01484	−0.00904	−0.03429	1.23225	0.40256	−0.19686	0.09016	0.80435	−1.54651	0.15583	−0.06477
3.10	0.03604	0.01484	−0.00904	−0.03429	1.23217	0.40255	−0.19685	0.09015	0.80438	−1.54651	0.15583	−0.06477

Table 8.7

h	$d^{(1)}_{13,2}$	$d^{(2)}_{13,2}$	$d^{(3)}_{13,2}$	$d^{(4)}_{13,2}$	$d^{(1)}_{13,3}$	$d^{(2)}_{13,3}$	$d^{(3)}_{13,3}$	$d^{(4)}_{13,3}$	$d^{(1)}_{13,4}$	$d^{(2)}_{13,4}$	$d^{(3)}_{13,4}$	$d^{(4)}_{13,4}$
1.05	−0.07097	1.01266	−4.16894	0.13756	0.03402	−0.07772	1.18714	−6.70637	−0.02065	0.03067	−0.08555	1.34427
1.10	−0.07000	1.04120	−4.15347	0.13680	0.03367	−0.07671	1.22430	−6.68247	−0.02046	0.03036	−0.08448	1.38987
1.15	−0.06920	1.06478	−4.14068	0.13617	0.03338	−0.07587	1.25502	−6.66272	−0.02030	0.03010	−0.08359	1.42755
1.20	−0.06854	1.08430	−4.13011	0.13565	0.03313	−0.07518	1.28042	−6.64638	−0.02017	0.02989	−0.08286	1.45872
1.25	−0.06799	1.10045	−4.12135	0.13522	0.03293	−0.07461	1.30146	−6.63285	−0.02006	0.02972	−0.08225	1.48454
1.30	−0.06754	1.11384	−4.11409	0.13486	0.03277	−0.07414	1.31890	−6.62163	−0.01997	0.02957	−0.08174	1.50593
1.35	−0.06716	1.12495	−4.10807	0.13456	0.03263	−0.07374	1.33336	−6.61233	−0.01989	0.02945	−0.08133	1.52367
1.40	−0.06685	1.13416	−4.10308	0.13432	0.03252	−0.07342	1.34536	−6.60462	−0.01983	0.02936	−0.08098	1.53839
1.45	−0.06659	1.14181	−4.09893	0.13411	0.03242	−0.07315	1.35532	−6.59821	−0.01978	0.02927	−0.08069	1.55061
1.50	−0.06638	1.14816	−4.09549	0.13394	0.03234	−0.07292	1.36359	−6.59290	−0.01974	0.02920	−0.08045	1.56076
1.55	−0.06620	1.15343	−4.09263	0.13380	0.03228	−0.07274	1.37045	−6.58848	−0.01970	0.02915	−0.08025	1.56918
1.60	−0.06605	1.15781	−4.09026	0.13368	0.03222	−0.07258	1.37616	−6.58481	−0.01967	0.02910	−0.08009	1.57618
1.65	−0.06593	1.16145	−4.08829	0.13359	0.03218	−0.07245	1.38090	−6.58176	−0.01965	0.02906	−0.07995	1.58200
1.70	−0.06582	1.16448	−4.08665	0.13351	0.03214	−0.07235	1.38484	−6.57923	−0.01963	0.02903	−0.07984	1.58684
1.75	−0.06574	1.16700	−4.08528	0.13344	0.03211	−0.07226	1.38812	−6.57712	−0.01961	0.02900	−0.07974	1.59086
1.80	−0.06567	1.16909	−4.08415	0.13338	0.03208	−0.07218	1.39084	−6.57537	−0.01960	0.02898	−0.07966	1.59420
1.85	−0.06561	1.17083	−4.08321	0.13334	0.03206	−0.07212	1.39311	−6.57391	−0.01958	0.02896	−0.07960	1.59698
1.90	−0.06556	1.17227	−4.08242	0.13330	0.03204	−0.07207	1.39499	−6.57270	−0.01957	0.02894	−0.07954	1.59929
1.95	−0.06552	1.17347	−4.08177	0.13327	0.03203	−0.07203	1.39655	−6.57170	−0.01957	0.02893	−0.07950	1.60121
2.00	−0.06548	1.17447	−4.08123	0.13324	0.03201	−0.07199	1.39786	−6.57086	−0.01956	0.02892	−0.07946	1.60280
2.05	−0.06546	1.17531	−4.08078	0.13322	0.03200	−0.07196	1.39894	−6.57016	−0.01955	0.02891	−0.07943	1.60413
2.10	−0.06543	1.17600	−4.08040	0.13320	0.03200	−0.07194	1.39984	−6.56958	−0.01955	0.02890	−0.07940	1.60524
2.15	−0.06541	1.17657	−4.08009	0.13318	0.03199	−0.07192	1.40059	−6.56910	−0.01955	0.02890	−0.07938	1.60616
2.20	−0.06540	1.17705	−4.07983	0.13317	0.03198	−0.07190	1.40121	−6.56870	−0.01954	0.02889	−0.07936	1.60692
2.25	−0.06538	1.17745	−4.07962	0.13316	0.03198	−0.07189	1.40173	−6.56837	−0.01954	0.02889	−0.07935	1.60756
2.30	−0.06537	1.17778	−4.07944	0.13315	0.03197	−0.07188	1.40216	−6.56809	−0.01954	0.02888	−0.07934	1.60809
2.40	−0.06535	1.17828	−4.07916	0.13314	0.03197	−0.07186	1.40282	−6.56767	−0.01953	0.02888	−0.07932	1.60889
2.50	−0.06534	1.17863	−4.07897	0.13313	0.03196	−0.07185	1.40327	−6.56738	−0.01953	0.02888	−0.07930	1.60945
2.60	−0.06533	1.17887	−4.07884	0.13312	0.03196	−0.07184	1.40359	−6.56717	−0.01953	0.02887	−0.07930	1.60983
2.70	−0.06533	1.17904	−4.07875	0.13312	0.03196	−0.07183	1.40380	−6.56703	−0.01953	0.02887	−0.07929	1.61010
2.80	−0.06533	1.17916	−4.07869	0.13312	0.03196	−0.07183	1.40395	−6.56694	−0.01953	0.02887	−0.07929	1.61029
2.90	−0.06532	1.17924	−4.07865	0.13311	0.03196	−0.07182	1.40406	−6.56687	−0.01953	0.02887	−0.07928	1.61041
3.00	−0.06532	1.17929	−4.07862	0.13311	0.03196	−0.07182	1.40413	−6.56682	−0.01953	0.02887	−0.07928	1.61050
3.10	−0.06532	1.17933	−4.07860	0.13311	0.03195	−0.07182	1.40418	−6.56679	−0.01953	0.02887	−0.07928	1.61056

Tables of hydrodynamic coefficients

Table 8.8

h	$d_{13,5}^{(1)}$	$d_{13,5}^{(2)}$	$d_{13,5}^{(3)}$	$d_{13,5}^{(4)}$	$d_{14,1}^{(1)}$	$d_{14,1}^{(2)}$	$d_{14,1}^{(3)}$	$d_{14,1}^{(4)}$	$d_{14,2}^{(1)}$	$d_{14,2}^{(2)}$	$d_{14,2}^{(3)}$	$d_{14,2}^{(4)}$
1.05	0.25667	0.03127	−0.00917	0.00416	−0.02934	0.07919	0.00527	−0.00319	0.01565	−0.00788	0.04285	−0.00068
1.10	0.25964	0.03120	−0.00917	0.00416	−0.02926	0.07919	0.00527	−0.00319	0.01562	−0.00788	0.04285	−0.00068
1.15	0.26180	0.03116	−0.00917	0.00416	−0.02919	0.07920	0.00527	−0.00319	0.01560	−0.00788	0.04285	−0.00068
1.20	0.26337	0.03114	−0.00917	0.00416	−0.02915	0.07920	0.00527	−0.00319	0.01558	−0.00788	0.04285	−0.00068
1.25	0.26452	0.03112	−0.00917	0.00416	−0.02912	0.07920	0.00527	−0.00319	0.01558	−0.00788	0.04285	−0.00068
1.30	0.26535	0.03111	−0.00917	0.00416	−0.02909	0.07920	0.00527	−0.00319	0.01557	−0.00788	0.04285	−0.00068
1.35	0.26596	0.03110	−0.00917	0.00416	−0.02907	0.07920	0.00527	−0.00319	0.01556	−0.00788	0.04285	−0.00068
1.40	0.26641	0.03110	−0.00917	0.00416	−0.02906	0.07920	0.00527	−0.00319	0.01556	−0.00788	0.04285	−0.00068
1.45	0.26674	0.03109	−0.00917	0.00416	−0.02905	0.07920	0.00527	−0.00319	0.01555	−0.00788	0.04285	−0.00068
1.50	0.26698	0.03109	−0.00917	0.00416	−0.02904	0.07920	0.00527	−0.00319	0.01555	−0.00788	0.04285	−0.00068
1.55	0.26715	0.03109	−0.00917	0.00416	−0.02904	0.07920	0.00527	−0.00319	0.01555	−0.00788	0.04285	−0.00068
1.60	0.26728	0.03109	−0.00917	0.00416	−0.02903	0.07920	0.00527	−0.00319	0.01555	−0.00788	0.04285	−0.00068
1.65	0.26738	0.03109	−0.00917	0.00416	−0.02903	0.07920	0.00527	−0.00319	0.01554	−0.00788	0.04285	−0.00068
1.70	0.26745	0.03109	−0.00917	0.00416	−0.02903	0.07920	0.00527	−0.00319	0.01554	−0.00788	0.04285	−0.00068
1.75	0.26750	0.03109	−0.00917	0.00416	−0.02903	0.07920	0.00527	−0.00319	0.01554	−0.00788	0.04285	−0.00068
1.80	0.26753	0.03109	−0.00917	0.00416	−0.02903	0.07920	0.00527	−0.00319	0.01554	−0.00788	0.04285	−0.00068
1.85	0.26756	0.03109	−0.00917	0.00416	−0.02903	0.07920	0.00527	−0.00319	0.01554	−0.00788	0.04285	−0.00068
1.90	0.26758	0.03109	−0.00917	0.00416	−0.02902	0.07920	0.00527	−0.00319	0.01554	−0.00788	0.04285	−0.00068
1.95	0.26760	0.03109	−0.00917	0.00416	−0.02902	0.07920	0.00527	−0.00319	0.01554	−0.00788	0.04285	−0.00068
2.00	0.26761	0.03109	−0.00917	0.00416	−0.02902	0.07920	0.00527	−0.00319	0.01554	−0.00788	0.04285	−0.00068
2.05	0.26762	0.03109	−0.00917	0.00416	−0.02902	0.07920	0.00527	−0.00319	0.01554	−0.00788	0.04285	−0.00068
2.10	0.26762	0.03109	−0.00917	0.00416	−0.02902	0.07920	0.00527	−0.00319	0.01554	−0.00788	0.04285	−0.00068
2.15	0.26763	0.03109	−0.00917	0.00416	−0.02902	0.07920	0.00527	−0.00319	0.01554	−0.00788	0.04285	−0.00068
2.20	0.26763	0.03109	−0.00917	0.00416	−0.02902	0.07920	0.00527	−0.00319	0.01554	−0.00788	0.04285	−0.00068
2.25	0.26763	0.03109	−0.00917	0.00416	−0.02902	0.07920	0.00527	−0.00319	0.01554	−0.00788	0.04285	−0.00068
2.30	0.26764	0.03109	−0.00917	0.00416	−0.02902	0.07920	0.00527	−0.00319	0.01554	−0.00788	0.04285	−0.00068
2.40	0.26764	0.03109	−0.00917	0.00416	−0.02902	0.07920	0.00527	−0.00319	0.01554	−0.00788	0.04285	−0.00068
2.50	0.26764	0.03109	−0.00917	0.00416	−0.02902	0.07920	0.00527	−0.00319	0.01554	−0.00788	0.04285	−0.00068
2.60	0.26764	0.03109	−0.00917	0.00416	−0.02902	0.07920	0.00527	−0.00319	0.01554	−0.00788	0.04285	−0.00068
2.70	0.26764	0.03109	−0.00917	0.00416	−0.02902	0.07920	0.00527	−0.00319	0.01554	−0.00788	0.04285	−0.00068
2.80	0.26764	0.03109	−0.00917	0.00416	−0.02902	0.07920	0.00527	−0.00319	0.01554	−0.00788	0.04285	−0.00068
2.90	0.26764	0.03109	−0.00917	0.00416	−0.02902	0.07920	0.00527	−0.00319	0.01554	−0.00788	0.04285	−0.00068
3.00	0.26764	0.03109	−0.00917	0.00416	−0.02902	0.07920	0.00527	−0.00319	0.01554	−0.00788	0.04285	−0.00068
3.10	0.26764	0.03109	−0.00917	0.00416	−0.02902	0.07920	0.00527	−0.00319	0.01554	−0.00788	0.04285	−0.00068

Table 8.9

h	$d_{14,3}^{(1)}$	$d_{14,3}^{(2)}$	$d_{14,3}^{(3)}$	$d_{14,3}^{(4)}$	$d_{14,4}^{(1)}$	$d_{14,4}^{(2)}$	$d_{14,4}^{(3)}$	$d_{14,4}^{(4)}$	$d_{15,1}^{(1)}$	$d_{15,1}^{(2)}$	$d_{15,1}^{(3)}$	$d_{15,1}^{(4)}$
1.05	−0.01002	0.00396	−0.00416	0.02830	−3.87540	−1.73540	0.18864	−0.07758	−0.48821	−7.39395	−4.34951	0.17562
1.10	−0.01000	0.00396	−0.00416	0.02830	−3.84228	−1.72315	0.18726	−0.07701	−0.48463	−7.34150	−4.31895	0.17437
1.15	−0.00999	0.00396	−0.00416	0.02830	−3.81532	−1.71299	0.18612	−0.07654	−0.48171	−7.29815	−4.29370	0.17333
1.20	−0.00999	0.00396	−0.00416	0.02830	−3.79330	−1.70457	0.18517	−0.07615	−0.47931	−7.26230	−4.27280	0.17247
1.25	−0.00998	0.00396	−0.00416	0.02830	−3.77528	−1.69759	0.18439	−0.07582	−0.47735	−7.23260	−4.25550	0.17176
1.30	−0.00998	0.00396	−0.00416	0.02830	−3.76050	−1.69180	0.18374	−0.07555	−0.47574	−7.20800	−4.24116	0.17117
1.35	−0.00997	0.00396	−0.00416	0.02830	−3.74835	−1.68700	0.18320	−0.07533	−0.47441	−7.18759	−4.22927	0.17069
1.40	−0.00997	0.00396	−0.00416	0.02830	−3.73836	−1.68300	0.18275	−0.07515	−0.47332	−7.17066	−4.21940	0.17028
1.45	−0.00997	0.00396	−0.00416	0.02830	−3.73013	−1.67969	0.18238	−0.07499	−0.47242	−7.15661	−4.21122	0.16995
1.50	−0.00997	0.00396	−0.00416	0.02830	−3.72334	−1.67693	0.18207	−0.07487	−0.47167	−7.14494	−4.20442	0.16967
1.55	−0.00997	0.00396	−0.00416	0.02830	−3.71774	−1.67465	0.18181	−0.07476	−0.47106	−7.13525	−4.19877	0.16943
1.60	−0.00997	0.00396	−0.00416	0.02830	−3.71310	−1.67274	0.18160	−0.07467	−0.47055	−7.12720	−4.19408	0.16924
1.65	−0.00997	0.00396	−0.00416	0.02830	−3.70927	−1.67116	0.18142	−0.07460	−0.47012	−7.12050	−4.19018	0.16908
1.70	−0.00996	0.00396	−0.00416	0.02830	−3.70610	−1.66985	0.18128	−0.07454	−0.46977	−7.11494	−4.18694	0.16895
1.75	−0.00996	0.00396	−0.00416	0.02830	−3.70347	−1.66876	0.18115	−0.07449	−0.46949	−7.11032	−4.18424	0.16884
1.80	−0.00996	0.00396	−0.00416	0.02830	−3.70129	−1.66785	0.18105	−0.07445	−0.46924	−7.10648	−4.18200	0.16875
1.85	−0.00996	0.00396	−0.00416	0.02830	−3.69949	−1.66709	0.18097	−0.07441	−0.46905	−7.10328	−4.18014	0.16867
1.90	−0.00996	0.00396	−0.00416	0.02830	−3.69799	−1.66647	0.18090	−0.07438	−0.46888	−7.10063	−4.17859	0.16861
1.95	−0.00996	0.00396	−0.00416	0.02830	−3.69675	−1.66594	0.18084	−0.07436	−0.46874	−7.09842	−4.17731	0.16855
2.00	−0.00996	0.00396	−0.00416	0.02830	−3.69572	−1.66551	0.18079	−0.07434	−0.46863	−7.09658	−4.17624	0.16851
2.05	−0.00996	0.00396	−0.00416	0.02830	−3.69486	−1.66515	0.18075	−0.07432	−0.46853	−7.09505	−4.17534	0.16847
2.10	−0.00996	0.00396	−0.00416	0.02830	−3.69415	−1.66485	0.18072	−0.07431	−0.46846	−7.09378	−4.17460	0.16844
2.15	−0.00996	0.00396	−0.00416	0.02830	−3.69356	−1.66460	0.18069	−0.07429	−0.46839	−7.09272	−4.17399	0.16842
2.20	−0.00996	0.00396	−0.00416	0.02830	−3.69307	−1.66439	0.18067	−0.07429	−0.46834	−7.09184	−4.17348	0.16840
2.25	−0.00996	0.00396	−0.00416	0.02830	−3.69267	−1.66422	0.18065	−0.07428	−0.46829	−7.09111	−4.17305	0.16838
2.30	−0.00996	0.00396	−0.00416	0.02830	−3.69233	−1.66407	0.18063	−0.07427	−0.46825	−7.09050	−4.17270	0.16836
2.40	−0.00996	0.00396	−0.00416	0.02830	−3.69181	−1.66385	0.18061	−0.07426	−0.46820	−7.08958	−4.17216	0.16834
2.50	−0.00996	0.00396	−0.00416	0.02830	−3.69146	−1.66370	0.18059	−0.07425	−0.46816	−7.08894	−4.17178	0.16833
2.60	−0.00996	0.00396	−0.00416	0.02830	−3.69121	−1.66360	0.18058	−0.07425	−0.46813	−7.08849	−4.17152	0.16832
2.70	−0.00996	0.00396	−0.00416	0.02830	−3.69104	−1.66353	0.18057	−0.07425	−0.46811	−7.08819	−4.17134	0.16831
2.80	−0.00996	0.00396	−0.00416	0.02830	−3.69092	−1.66348	0.18056	−0.07424	−0.46810	−7.08797	−4.17122	0.16831
2.90	−0.00996	0.00396	−0.00416	0.02830	−3.69084	−1.66344	0.18056	−0.07424	−0.46809	−7.08783	−4.17114	0.16830
3.00	−0.00996	0.00396	−0.00416	0.02830	−3.69078	−1.66342	0.18056	−0.07424	−0.46808	−7.08773	−4.17108	0.16830
3.10	−0.00996	0.00396	−0.00416	0.02830	−3.69075	−1.66340	0.18055	−0.07424	−0.46808	−7.08766	−4.17104	0.16830

Tables of hydrodynamic coefficients 131

Table 8.10

h	$d_{15,2}^{(1)}$	$d_{15,2}^{(2)}$	$d_{15,2}^{(3)}$	$d_{15,2}^{(4)}$	$d_{15,3}^{(1)}$	$d_{15,3}^{(2)}$	$d_{15,3}^{(3)}$	$d_{15,3}^{(4)}$	$d_{15,4}^{(1)}$	$d_{15,4}^{(2)}$	$d_{15,4}^{(3)}$	$d_{15,4}^{(4)}$
1.05	0.24399	−0.37619	−10.03128	−6.82125	−0.16499	0.13946	−0.36825	−12.52179	0.81167	0.37236	0.23423	0.17085
1.10	0.24220	−0.37360	−9.96058	−6.77338	−0.16378	0.13851	−0.36569	−12.43372	0.83300	0.37301	0.23425	0.17085
1.15	0.24075	−0.37147	−9.90216	−6.73383	−0.16279	0.13772	−0.36358	−12.36094	0.85288	0.37351	0.23427	0.17085
1.20	0.23956	−0.36970	−9.85382	−6.70110	−0.16198	0.13707	−0.36183	−12.30072	0.87138	0.37389	0.23428	0.17085
1.25	0.23858	−0.36824	−9.81380	−6.67400	−0.16131	0.13653	−0.36039	−12.25086	0.88856	0.37418	0.23428	0.17085
1.30	0.23778	−0.36702	−9.78063	−6.65154	−0.16077	0.13609	−0.35919	−12.20954	0.90449	0.37440	0.23429	0.17085
1.35	0.23712	−0.36602	−9.75312	−6.63292	−0.16032	0.13572	−0.35820	−12.17527	0.91925	0.37457	0.23429	0.17085
1.40	0.23657	−0.36518	−9.73030	−6.61747	−0.15995	0.13541	−0.35737	−12.14684	0.93290	0.37470	0.23429	0.17085
1.45	0.23612	−0.36449	−9.71135	−6.60464	−0.15965	0.13516	−0.35669	−12.12324	0.94551	0.37480	0.23429	0.17085
1.50	0.23575	−0.36392	−9.69563	−6.59399	−0.15939	0.13494	−0.35612	−12.10364	0.95716	0.37488	0.23429	0.17085
1.55	0.23545	−0.36344	−9.68256	−6.58515	−0.15919	0.13477	−0.35565	−12.08737	0.96789	0.37494	0.23429	0.17085
1.60	0.23519	−0.36304	−9.67171	−6.57780	−0.15901	0.13462	−0.35525	−12.07384	0.97778	0.37498	0.23429	0.17085
1.65	0.23498	−0.36271	−9.66269	−6.57169	−0.15887	0.13450	−0.35493	−12.06261	0.98689	0.37502	0.23429	0.17085
1.70	0.23481	−0.36244	−9.65519	−6.56662	−0.15875	0.13440	−0.35466	−12.05327	0.99526	0.37505	0.23429	0.17085
1.75	0.23466	−0.36221	−9.64896	−6.56240	−0.15866	0.13432	−0.35443	−12.04550	1.00296	0.37507	0.23429	0.17085
1.80	0.23454	−0.36202	−9.64378	−6.55889	−0.15857	0.13425	−0.35424	−12.03905	1.01002	0.37508	0.23429	0.17085
1.85	0.23445	−0.36186	−9.63947	−6.55597	−0.15851	0.13419	−0.35409	−12.03368	1.01651	0.37509	0.23429	0.17085
1.90	0.23436	−0.36173	−9.63589	−6.55355	−0.15845	0.13414	−0.35396	−12.02922	1.02247	0.37510	0.23429	0.17085
1.95	0.23430	−0.36162	−9.63291	−6.55153	−0.15840	0.13410	−0.35385	−12.02551	1.02793	0.37511	0.23429	0.17085
2.00	0.23424	−0.36153	−9.63043	−6.54986	−0.15837	0.13407	−0.35376	−12.02243	1.03293	0.37512	0.23429	0.17085
2.05	0.23419	−0.36146	−9.62837	−6.54846	−0.15833	0.13404	−0.35369	−12.01986	1.03752	0.37512	0.23429	0.17085
2.10	0.23415	−0.36139	−9.62666	−6.54730	−0.15831	0.13402	−0.35362	−12.01773	1.04172	0.37512	0.23429	0.17085
2.15	0.23412	−0.36134	−9.62523	−6.54634	−0.15829	0.13400	−0.35357	−12.01595	1.04556	0.37513	0.23429	0.17085
2.20	0.23409	−0.36130	−9.62405	−6.54553	−0.15827	0.13398	−0.35353	−12.01447	1.04908	0.37513	0.23429	0.17085
2.25	0.23407	−0.36126	−9.62306	−6.54487	−0.15825	0.13397	−0.35349	−12.01325	1.05230	0.37513	0.23429	0.17085
2.30	0.23405	−0.36123	−9.62224	−6.54431	−0.15824	0.13396	−0.35347	−12.01222	1.05524	0.37513	0.23429	0.17085
2.40	0.23402	−0.36119	−9.62099	−6.54347	−0.15822	0.13394	−0.35342	−12.01067	1.06039	0.37513	0.23429	0.17085
2.50	0.23400	−0.36116	−9.62013	−6.54288	−0.15821	0.13393	−0.35339	−12.00959	1.06470	0.37513	0.23429	0.17085
2.60	0.23399	−0.36113	−9.61953	−6.54248	−0.15820	0.13392	−0.35337	−12.00885	1.06830	0.37513	0.23429	0.17085
2.70	0.23398	−0.36112	−9.61912	−6.54220	−0.15819	0.13392	−0.35335	−12.00833	1.07130	0.37513	0.23429	0.17085
2.80	0.23397	−0.36111	−9.61883	−6.54200	−0.15819	0.13391	−0.35334	−12.00798	1.07380	0.37513	0.23429	0.17085
2.90	0.23397	−0.36110	−9.61863	−6.54187	−0.15818	0.13391	−0.35333	−12.00773	1.07588	0.37513	0.23429	0.17085
3.00	0.23397	−0.36110	−9.61850	−6.54178	−0.15818	0.13391	−0.35333	−12.00756	1.07762	0.37513	0.23429	0.17085
3.10	0.23396	−0.36109	−9.61840	−6.54171	−0.15818	0.13391	−0.35333	−12.00744	1.07907	0.37513	0.23429	0.17085

Table 8.11

h	$d_{16,2}$	$d_{16,3}$	$d_{16,4}$	$d_{17,2}$	$d_{17,3}$	$d_{17,4}$	$d_{18,2}$	$d_{18,3}$	$d_{18,4}$	$d_{19,2}$	$d_{19,3}$	$d_{19,4}$
1.05	−0.11741	0.01572	−0.00966	−0.50027	0.12018	−0.06049	−4.08633	0.32527	−0.19563	−4.47672	0.80535	−0.49341
1.10	−0.09908	0.01495	−0.00933	−0.49234	0.11955	−0.06020	−3.96474	0.31695	−0.19085	−4.39060	0.79822	−0.48932
1.15	−0.08433	0.01433	−0.00905	−0.48597	0.11905	−0.05996	−3.86606	0.31020	−0.18697	−4.32050	0.79237	−0.48596
1.20	−0.07241	0.01382	−0.00883	−0.48084	0.11865	−0.05977	−3.78567	0.30469	−0.18381	−4.26325	0.78755	−0.48318
1.25	−0.06272	0.01340	−0.00864	−0.47669	0.11832	−0.05961	−3.71997	0.30019	−0.18122	−4.21636	0.78357	−0.48089
1.30	−0.05483	0.01307	−0.00849	−0.47333	0.11805	−0.05948	−3.66613	0.29650	−0.17910	−4.17787	0.78029	−0.47900
1.35	−0.04838	0.01279	−0.00837	−0.47059	0.11783	−0.05938	−3.62190	0.29347	−0.17736	−4.14620	0.77758	−0.47743
1.40	−0.04310	0.01256	−0.00827	−0.46836	0.11765	−0.05929	−3.58550	0.29098	−0.17593	−4.12011	0.77533	−0.47614
1.45	−0.03876	0.01237	−0.00819	−0.46654	0.11751	−0.05922	−3.55550	0.28893	−0.17475	−4.09857	0.77347	−0.47506
1.50	−0.03519	0.01222	−0.00812	−0.46504	0.11739	−0.05916	−3.53074	0.28724	−0.17378	−4.08078	0.77194	−0.47417
1.55	−0.03225	0.01209	−0.00806	−0.46382	0.11729	−0.05912	−3.51027	0.28584	−0.17297	−4.06606	0.77066	−0.47343
1.60	−0.02983	0.01199	−0.00801	−0.46281	0.11721	−0.05908	−3.49334	0.28469	−0.17231	−4.05388	0.76960	−0.47282
1.65	−0.02783	0.01190	−0.00798	−0.46198	0.11714	−0.05904	−3.47932	0.28373	−0.17176	−4.04378	0.76872	−0.47231
1.70	−0.02617	0.01183	−0.00794	−0.46130	0.11708	−0.05902	−3.46771	0.28294	−0.17130	−4.03542	0.76799	−0.47189
1.75	−0.02480	0.01177	−0.00792	−0.46074	0.11704	−0.05899	−3.45809	0.28228	−0.17092	−4.02848	0.76739	−0.47154
1.80	−0.02367	0.01172	−0.00790	−0.46027	0.11700	−0.05898	−3.45010	0.28174	−0.17061	−4.02272	0.76689	−0.47125
1.85	−0.02273	0.01168	−0.00788	−0.45989	0.11697	−0.05896	−3.44348	0.28129	−0.17035	−4.01795	0.76647	−0.47100
1.90	−0.02195	0.01165	−0.00786	−0.45957	0.11694	−0.05895	−3.43798	0.28092	−0.17014	−4.01398	0.76612	−0.47080
1.95	−0.02130	0.01162	−0.00785	−0.45931	0.11692	−0.05894	−3.43341	0.28061	−0.16996	−4.01068	0.76583	−0.47064
2.00	−0.02076	0.01160	−0.00784	−0.45909	0.11691	−0.05893	−3.42962	0.28035	−0.16981	−4.00795	0.76559	−0.47050
2.05	−0.02032	0.01158	−0.00783	−0.45891	0.11689	−0.05892	−3.42647	0.28013	−0.16969	−4.00567	0.76539	−0.47038
2.10	−0.01995	0.01156	−0.00782	−0.45876	0.11688	−0.05892	−3.42386	0.27996	−0.16958	−4.00378	0.76523	−0.47029
2.15	−0.01964	0.01155	−0.00782	−0.45864	0.11687	−0.05891	−3.42168	0.27981	−0.16950	−4.00221	0.76509	−0.47021
2.20	−0.01939	0.01154	−0.00781	−0.45853	0.11686	−0.05891	−3.41987	0.27969	−0.16943	−4.00091	0.76498	−0.47014
2.25	−0.01917	0.01153	−0.00781	−0.45845	0.11685	−0.05890	−3.41837	0.27958	−0.16937	−3.99982	0.76488	−0.47008
2.30	−0.01900	0.01152	−0.00781	−0.45838	0.11685	−0.05890	−3.41712	0.27950	−0.16932	−3.99892	0.76480	−0.47004
2.40	−0.01873	0.01151	−0.00780	−0.45827	0.11684	−0.05890	−3.41522	0.27937	−0.16925	−3.99755	0.76468	−0.46997
2.50	−0.01855	0.01150	−0.00780	−0.45820	0.11683	−0.05889	−3.41391	0.27928	−0.16920	−3.99660	0.76460	−0.46992
2.60	−0.01842	0.01150	−0.00780	−0.45815	0.11683	−0.05889	−3.41300	0.27922	−0.16916	−3.99594	0.76454	−0.46989
2.70	−0.01833	0.01149	−0.00779	−0.45811	0.11683	−0.05889	−3.41237	0.27918	−0.16914	−3.99549	0.76450	−0.46986
2.80	−0.01827	0.01149	−0.00779	−0.45809	0.11682	−0.05889	−3.41193	0.27915	−0.16912	−3.99517	0.76447	−0.46985
2.90	−0.01823	0.01149	−0.00779	−0.45807	0.11682	−0.05889	−3.41163	0.27913	−0.16911	−3.99495	0.76445	−0.46984
3.00	−0.01820	0.01149	−0.00779	−0.45806	0.11682	−0.05889	−3.41143	0.27911	−0.16910	−3.99480	0.76444	−0.46983
3.10	−0.01818	0.01149	−0.00779	−0.45805	0.11682	−0.05889	−3.41128	0.27910	−0.16909	−3.99470	0.76443	−0.46982

Tables of hydrodynamic coefficients 133

Table 8.12

h	$d_{20,2}^{(1)}$	$d_{20,2}^{(2)}$	$d_{20,2}^{(3)}$	$d_{20,2}^{(4)}$	$d_{20,3}^{(1)}$	$d_{20,3}^{(2)}$	$d_{20,3}^{(3)}$	$d_{20,3}^{(4)}$	$d_{20,4}^{(1)}$	$d_{20,4}^{(2)}$	$d_{20,4}^{(3)}$	$d_{20,4}^{(4)}$
1.05	0.19959	5.25295	0.10010	−0.02706	−0.06083	0.09486	7.93455	0.11651	0.03173	−0.04283	−0.02893	10.52344
1.10	0.19097	5.23782	0.09955	−0.02689	−0.06006	0.07661	7.91038	0.11588	0.03141	−0.04230	−0.05597	10.49063
1.15	0.18385	5.22531	0.09910	−0.02675	−0.05941	0.06154	7.89040	0.11537	0.03114	−0.04186	−0.07832	10.46352
1.20	0.17796	5.21495	0.09872	−0.02664	−0.05888	0.04906	7.87387	0.11494	0.03091	−0.04150	−0.09681	10.44110
1.25	0.17309	5.20638	0.09841	−0.02654	−0.05844	0.03873	7.86018	0.11459	0.03073	−0.04120	−0.11212	10.42252
1.30	0.16905	5.19926	0.09815	−0.02646	−0.05808	0.03017	7.84884	0.11429	0.03058	−0.04095	−0.12481	10.40713
1.35	0.16570	5.19337	0.09794	−0.02640	−0.05778	0.02307	7.83943	0.11405	0.03045	−0.04074	−0.13533	10.39437
1.40	0.16292	5.18847	0.09776	−0.02634	−0.05753	0.01718	7.83163	0.11385	0.03035	−0.04057	−0.14406	10.38378
1.45	0.16061	5.18441	0.09762	−0.02630	−0.05732	0.01229	7.82515	0.11368	0.03026	−0.04043	−0.15130	10.37499
1.50	0.15870	5.18104	0.09749	−0.02626	−0.05714	0.00824	7.81977	0.11354	0.03019	−0.04031	−0.15732	10.36769
1.55	0.15711	5.17823	0.09739	−0.02623	−0.05700	0.00486	7.81530	0.11343	0.03013	−0.04021	−0.16232	10.36163
1.60	0.15578	5.17591	0.09731	−0.02621	−0.05688	0.00206	7.81159	0.11333	0.03008	−0.04013	−0.16647	10.35659
1.65	0.15469	5.17397	0.09724	−0.02618	−0.05678	−0.00027	7.80850	0.11325	0.03004	−0.04007	−0.16992	10.35240
1.70	0.15377	5.17236	0.09718	−0.02617	−0.05670	−0.00220	7.80594	0.11319	0.03000	−0.04001	−0.17279	10.34893
1.75	0.15301	5.17103	0.09713	−0.02615	−0.05663	−0.00381	7.80381	0.11313	0.02997	−0.03996	−0.17517	10.34603
1.80	0.15238	5.16991	0.09709	−0.02614	−0.05657	−0.00514	7.80204	0.11309	0.02995	−0.03992	−0.17715	10.34363
1.85	0.15186	5.16899	0.09706	−0.02613	−0.05653	−0.00626	7.80056	0.11305	0.02993	−0.03989	−0.17880	10.34163
1.90	0.15142	5.16822	0.09703	−0.02612	−0.05649	−0.00718	7.79934	0.11302	0.02991	−0.03986	−0.18017	10.33997
1.95	0.15106	5.16758	0.09701	−0.02611	−0.05646	−0.00795	7.79832	0.11299	0.02990	−0.03984	−0.18131	10.33859
2.00	0.15076	5.16705	0.09699	−0.02611	−0.05643	−0.00859	7.79747	0.11297	0.02989	−0.03982	−0.18226	10.33744
2.05	0.15051	5.16661	0.09697	−0.02610	−0.05641	−0.00912	7.79677	0.11295	0.02988	−0.03981	−0.18304	10.33648
2.10	0.15030	5.16624	0.09696	−0.02610	−0.05639	−0.00956	7.79618	0.11294	0.02987	−0.03980	−0.18370	10.33569
2.15	0.15013	5.16594	0.09695	−0.02610	−0.05637	−0.00993	7.79570	0.11292	0.02986	−0.03978	−0.18424	10.33503
2.20	0.14998	5.16568	0.09694	−0.02609	−0.05636	−0.01024	7.79529	0.11291	0.02986	−0.03978	−0.18470	10.33448
2.25	0.14986	5.16547	0.09693	−0.02609	−0.05635	−0.01049	7.79495	0.11290	0.02985	−0.03977	−0.18507	10.33402
2.30	0.14976	5.16530	0.09692	−0.02609	−0.05634	−0.01070	7.79467	0.11290	0.02985	−0.03976	−0.18539	10.33364
2.40	0.14961	5.16503	0.09691	−0.02608	−0.05632	−0.01102	7.79425	0.11289	0.02984	−0.03975	−0.18587	10.33306
2.50	0.14951	5.16484	0.09691	−0.02608	−0.05631	−0.01125	7.79395	0.11288	0.02984	−0.03975	−0.18620	10.33266
2.60	0.14943	5.16471	0.09690	−0.02608	−0.05631	−0.01140	7.79375	0.11287	0.02984	−0.03974	−0.18642	10.33238
2.70	0.14938	5.16463	0.09690	−0.02608	−0.05630	−0.01151	7.79361	0.11287	0.02984	−0.03974	−0.18658	10.33219
2.80	0.14935	5.16456	0.09690	−0.02608	−0.05630	−0.01158	7.79351	0.11287	0.02983	−0.03974	−0.18669	10.33206
2.90	0.14932	5.16452	0.09690	−0.02608	−0.05630	−0.01163	7.79344	0.11286	0.02983	−0.03973	−0.18677	10.33196
3.00	0.14931	5.16449	0.09689	−0.02608	−0.05630	−0.01167	7.79339	0.11286	0.02983	−0.03973	−0.18682	10.33190
3.10	0.14929	5.16447	0.09689	−0.02608	−0.05630	−0.01169	7.79336	0.11286	0.02983	−0.03973	−0.18686	10.33186

Table 8.13

h	$d_{21,2}^{(1)}$	$d_{21,2}^{(2)}$	$d_{21,2}^{(3)}$	$d_{21,2}^{(4)}$	$d_{21,3}^{(1)}$	$d_{21,3}^{(2)}$	$d_{21,3}^{(3)}$	$d_{21,3}^{(4)}$	$d_{21,4}^{(1)}$	$d_{21,4}^{(2)}$	$d_{21,4}^{(3)}$	$d_{21,4}^{(4)}$
1.05	0.06021	0.06478	−0.00372	0.00131	−0.02574	−0.00088	0.03524	−0.00222	0.01470	−0.00450	−0.00623	0.02363
1.10	0.05926	0.06482	−0.00372	0.00131	−0.02567	−0.00088	0.03524	−0.00222	0.01468	−0.00450	−0.00623	0.02363
1.15	0.05857	0.06484	−0.00372	0.00131	−0.02562	−0.00088	0.03524	−0.00222	0.01466	−0.00450	−0.00623	0.02363
1.20	0.05805	0.06485	−0.00372	0.00131	−0.02559	−0.00088	0.03524	−0.00222	0.01464	−0.00450	−0.00623	0.02363
1.25	0.05768	0.06486	−0.00372	0.00131	−0.02556	−0.00088	0.03524	−0.00222	0.01463	−0.00450	−0.00623	0.02363
1.30	0.05740	0.06486	−0.00372	0.00131	−0.02554	−0.00088	0.03524	−0.00222	0.01463	−0.00450	−0.00623	0.02363
1.35	0.05720	0.06487	−0.00372	0.00131	−0.02553	−0.00088	0.03524	−0.00222	0.01462	−0.00450	−0.00623	0.02363
1.40	0.05705	0.06487	−0.00372	0.00131	−0.02552	−0.00088	0.03524	−0.00222	0.01462	−0.00450	−0.00623	0.02363
1.45	0.05693	0.06487	−0.00372	0.00131	−0.02551	−0.00088	0.03524	−0.00222	0.01461	−0.00450	−0.00623	0.02363
1.50	0.05685	0.06487	−0.00372	0.00131	−0.02551	−0.00088	0.03524	−0.00222	0.01461	−0.00450	−0.00623	0.02363
1.55	0.05679	0.06487	−0.00372	0.00131	−0.02550	−0.00088	0.03524	−0.00222	0.01461	−0.00450	−0.00623	0.02363
1.60	0.05675	0.06487	−0.00372	0.00131	−0.02550	−0.00088	0.03524	−0.00222	0.01461	−0.00450	−0.00623	0.02363
1.65	0.05672	0.06487	−0.00372	0.00131	−0.02550	−0.00088	0.03524	−0.00222	0.01461	−0.00450	−0.00623	0.02363
1.70	0.05669	0.06487	−0.00372	0.00131	−0.02550	−0.00088	0.03524	−0.00222	0.01461	−0.00450	−0.00623	0.02363
1.75	0.05668	0.06487	−0.00372	0.00131	−0.02549	−0.00088	0.03524	−0.00222	0.01461	−0.00450	−0.00623	0.02363
1.80	0.05666	0.06487	−0.00372	0.00131	−0.02549	−0.00088	0.03524	−0.00222	0.01461	−0.00450	−0.00623	0.02363
1.85	0.05665	0.06487	−0.00372	0.00131	−0.02549	−0.00088	0.03524	−0.00222	0.01461	−0.00450	−0.00623	0.02363
1.90	0.05665	0.06487	−0.00372	0.00131	−0.02549	−0.00088	0.03524	−0.00222	0.01461	−0.00450	−0.00623	0.02363
1.95	0.05664	0.06487	−0.00372	0.00131	−0.02549	−0.00088	0.03524	−0.00222	0.01461	−0.00450	−0.00623	0.02363
2.00	0.05664	0.06487	−0.00372	0.00131	−0.02549	−0.00088	0.03524	−0.00222	0.01461	−0.00450	−0.00623	0.02363
2.05	0.05663	0.06487	−0.00372	0.00131	−0.02549	−0.00088	0.03524	−0.00222	0.01461	−0.00450	−0.00623	0.02363
2.10	0.05663	0.06487	−0.00372	0.00131	−0.02549	−0.00088	0.03524	−0.00222	0.01461	−0.00450	−0.00623	0.02363
2.15	0.05663	0.06487	−0.00372	0.00131	−0.02549	−0.00088	0.03524	−0.00222	0.01461	−0.00450	−0.00623	0.02363
2.20	0.05663	0.06487	−0.00372	0.00131	−0.02549	−0.00088	0.03524	−0.00222	0.01461	−0.00450	−0.00623	0.02363
2.25	0.05663	0.06487	−0.00372	0.00131	−0.02549	−0.00088	0.03524	−0.00222	0.01461	−0.00450	−0.00623	0.02363
2.30	0.05663	0.06487	−0.00372	0.00131	−0.02549	−0.00088	0.03524	−0.00222	0.01461	−0.00450	−0.00623	0.02363
2.40	0.05663	0.06487	−0.00372	0.00131	−0.02549	−0.00088	0.03524	−0.00222	0.01461	−0.00450	−0.00623	0.02363
2.50	0.05663	0.06487	−0.00372	0.00131	−0.02549	−0.00088	0.03524	−0.00222	0.01461	−0.00450	−0.00623	0.02363
2.60	0.05663	0.06487	−0.00372	0.00131	−0.02549	−0.00088	0.03524	−0.00222	0.01461	−0.00450	−0.00623	0.02363
2.70	0.05663	0.06487	−0.00372	0.00131	−0.02549	−0.00088	0.03524	−0.00222	0.01461	−0.00450	−0.00623	0.02363
2.80	0.05663	0.06487	−0.00372	0.00131	−0.02549	−0.00088	0.03524	−0.00222	0.01461	−0.00450	−0.00623	0.02363
2.90	0.05663	0.06487	−0.00372	0.00131	−0.02549	−0.00088	0.03524	−0.00222	0.01461	−0.00450	−0.00623	0.02363
3.00	0.05663	0.06487	−0.00372	0.00131	−0.02549	−0.00088	0.03524	−0.00222	0.01461	−0.00450	−0.00623	0.02363
3.10	0.05663	0.06487	−0.00372	0.00131	−0.02549	−0.00088	0.03524	−0.00222	0.01461	−0.00450	−0.00623	0.02363

Table 8.14

h	$d_{22,2}^{(1)}$	$d_{22,2}^{(2)}$	$d_{22,2}^{(3)}$	$d_{22,2}^{(4)}$	$d_{22,3}^{(1)}$	$d_{22,3}^{(2)}$	$d_{22,3}^{(3)}$	$d_{22,3}^{(4)}$	$d_{22,4}^{(1)}$	$d_{22,4}^{(2)}$	$d_{22,4}^{(3)}$	$d_{22,4}^{(4)}$
1.05	2.78707	4.82892	0.12064	−0.03270	−0.41534	5.19106	7.29465	0.14837	0.23004	−0.20256	7.62413	9.70557
1.10	2.76572	4.79555	0.11977	−0.03246	−0.41231	5.15462	7.24370	0.14731	0.22836	−0.20117	7.57058	9.63763
1.15	2.74828	4.76796	0.11905	−0.03227	−0.40983	5.12451	7.20160	0.14644	0.22699	−0.20002	7.52633	9.58148
1.20	2.73401	4.74512	0.11845	−0.03210	−0.40779	5.09960	7.16677	0.14572	0.22587	−0.19907	7.48972	9.53503
1.25	2.72231	4.72620	0.11796	−0.03197	−0.40613	5.07897	7.13793	0.14512	0.22494	−0.19829	7.45940	9.49656
1.30	2.71270	4.71052	0.11755	−0.03186	−0.40476	5.06187	7.11403	0.14462	0.22419	−0.19764	7.43428	9.46468
1.35	2.70479	4.69752	0.11721	−0.03176	−0.40363	5.04770	7.09421	0.14421	0.22357	−0.19709	7.41344	9.43824
1.40	2.69827	4.68673	0.11692	−0.03169	−0.40270	5.03594	7.07776	0.14387	0.22305	−0.19665	7.39616	9.41631
1.45	2.69289	4.67777	0.11669	−0.03162	−0.40194	5.02617	7.06411	0.14359	0.22263	−0.19627	7.38181	9.39810
1.50	2.68845	4.67033	0.11650	−0.03157	−0.40131	5.01807	7.05277	0.14335	0.22228	−0.19596	7.36989	9.38298
1.55	2.68478	4.66416	0.11634	−0.03152	−0.40078	5.01133	7.04336	0.14316	0.22199	−0.19571	7.36000	9.37043
1.60	2.68175	4.65902	0.11620	−0.03149	−0.40035	5.00574	7.03554	0.14299	0.22175	−0.19549	7.35178	9.36000
1.65	2.67924	4.65476	0.11609	−0.03146	−0.39999	5.00109	7.02904	0.14286	0.22155	−0.19532	7.34494	9.35133
1.70	2.67715	4.65121	0.11600	−0.03143	−0.39969	4.99723	7.02364	0.14275	0.22139	−0.19517	7.33927	9.34412
1.75	2.67543	4.64827	0.11592	−0.03141	−0.39945	4.99401	7.01915	0.14265	0.22125	−0.19505	7.33454	9.33813
1.80	2.67400	4.64582	0.11586	−0.03139	−0.39924	4.99134	7.01541	0.14258	0.22114	−0.19494	7.33062	9.33315
1.85	2.67281	4.64378	0.11580	−0.03138	−0.39908	4.98912	7.01231	0.14251	0.22105	−0.19486	7.32736	9.32901
1.90	2.67183	4.64208	0.11576	−0.03137	−0.39893	4.98728	7.00973	0.14246	0.22097	−0.19479	7.32465	9.32557
1.95	2.67101	4.64068	0.11572	−0.03136	−0.39882	4.98574	7.00758	0.14241	0.22090	−0.19473	7.32239	9.32271
2.00	2.67033	4.63950	0.11569	−0.03135	−0.39872	4.98447	7.00580	0.14238	0.22085	−0.19468	7.32051	9.32033
2.05	2.66977	4.63853	0.11567	−0.03134	−0.39864	4.98341	7.00431	0.14235	0.22081	−0.19464	7.31895	9.31835
2.10	2.66930	4.63772	0.11565	−0.03134	−0.39857	4.98252	7.00308	0.14232	0.22077	−0.19461	7.31766	9.31670
2.15	2.66891	4.63705	0.11563	−0.03133	−0.39852	4.98179	7.00205	0.14230	0.22074	−0.19458	7.31658	9.31533
2.20	2.66859	4.63649	0.11561	−0.03133	−0.39847	4.98118	7.00120	0.14228	0.22071	−0.19456	7.31568	9.31420
2.25	2.66832	4.63602	0.11560	−0.03132	−0.39843	4.98067	7.00049	0.14227	0.22069	−0.19454	7.31493	9.31325
2.30	2.66810	4.63563	0.11559	−0.03132	−0.39840	4.98025	6.99990	0.14225	0.22067	−0.19452	7.31431	9.31246
2.40	2.66776	4.63504	0.11558	−0.03132	−0.39835	4.97960	6.99900	0.14224	0.22065	−0.19450	7.31337	9.31126
2.50	2.66753	4.63463	0.11556	−0.03131	−0.39832	4.97916	6.99837	0.14222	0.22063	−0.19448	7.31271	9.31043
2.60	2.66736	4.63435	0.11556	−0.03131	−0.39830	4.97885	6.99794	0.14221	0.22062	−0.19447	7.31226	9.30985
2.70	2.66725	4.63415	0.11555	−0.03131	−0.39828	4.97864	6.99764	0.14221	0.22061	−0.19446	7.31195	9.30946
2.80	2.66717	4.63402	0.11555	−0.03131	−0.39827	4.97849	6.99744	0.14220	0.22060	−0.19445	7.31173	9.30918
2.90	2.66712	4.63393	0.11555	−0.03131	−0.39826	4.97839	6.99730	0.14220	0.22060	−0.19445	7.31158	9.30899
3.00	2.66708	4.63386	0.11554	−0.03131	−0.39826	4.97832	6.99720	0.14220	0.22059	−0.19445	7.31147	9.30886
3.10	2.66706	4.63382	0.11554	−0.03131	−0.39825	4.97827	6.99713	0.14220	0.22059	−0.19445	7.31140	9.30877

Table 8.15

h	$d_{23,2}^{(1)}$	$d_{23,2}^{(2)}$	$d_{23,2}^{(3)}$	$d_{23,2}^{(4)}$	$d_{23,3}^{(1)}$	$d_{23,3}^{(2)}$	$d_{23,3}^{(3)}$	$d_{23,3}^{(4)}$	$d_{23,4}^{(1)}$	$d_{23,4}^{(2)}$	$d_{23,4}^{(3)}$	$d_{23,4}^{(4)}$
1.05	0.46969	−7.50999	0.06961	−0.03088	−0.02170	0.81306	−12.54297	0.03928	0.00523	−0.03720	1.12818	−17.56223
1.10	0.50308	−7.48472	0.06922	−0.03068	−0.02140	0.86276	−12.50119	0.03907	0.00517	−0.03673	1.19435	−17.50394
1.15	0.53065	−7.46382	0.06890	−0.03053	−0.02115	0.90383	−12.46665	0.03890	0.00513	−0.03633	1.24904	−17.45577
1.20	0.55346	−7.44653	0.06864	−0.03039	−0.02095	0.93780	−12.43809	0.03875	0.00509	−0.03601	1.29428	−17.41591
1.25	0.57234	−7.43221	0.06842	−0.03029	−0.02078	0.96594	−12.41443	0.03863	0.00506	−0.03574	1.33175	−17.38291
1.30	0.58798	−7.42033	0.06824	−0.03020	−0.02064	0.98925	−12.39483	0.03853	0.00503	−0.03551	1.36280	−17.35556
1.35	0.60095	−7.41049	0.06809	−0.03012	−0.02052	1.00859	−12.37857	0.03845	0.00501	−0.03533	1.38855	−17.33287
1.40	0.61171	−7.40231	0.06796	−0.03006	−0.02043	1.02463	−12.36508	0.03838	0.00499	−0.03517	1.40991	−17.31406
1.45	0.62064	−7.39553	0.06786	−0.03001	−0.02035	1.03795	−12.35388	0.03832	0.00498	−0.03504	1.42764	−17.29844
1.50	0.62806	−7.38990	0.06777	−0.02996	−0.02028	1.04900	−12.34459	0.03828	0.00497	−0.03494	1.44236	−17.28547
1.55	0.63422	−7.38522	0.06770	−0.02993	−0.02023	1.05819	−12.33686	0.03824	0.00496	−0.03485	1.45459	−17.27469
1.60	0.63933	−7.38133	0.06764	−0.02990	−0.02018	1.06582	−12.33045	0.03820	0.00495	−0.03478	1.46476	−17.26574
1.65	0.64359	−7.37810	0.06759	−0.02988	−0.02014	1.07216	−12.32512	0.03818	0.00494	−0.03471	1.47320	−17.25831
1.70	0.64712	−7.37542	0.06755	−0.02985	−0.02011	1.07743	−12.32069	0.03815	0.00493	−0.03466	1.48022	−17.25212
1.75	0.65006	−7.37319	0.06752	−0.02984	−0.02008	1.08181	−12.31700	0.03814	0.00493	−0.03462	1.48605	−17.24699
1.80	0.65250	−7.37133	0.06749	−0.02982	−0.02006	1.08545	−12.31394	0.03812	0.00493	−0.03459	1.49090	−17.24271
1.85	0.65453	−7.36979	0.06747	−0.02981	−0.02004	1.08848	−12.31140	0.03811	0.00492	−0.03456	1.49493	−17.23916
1.90	0.65622	−7.36851	0.06745	−0.02980	−0.02003	1.09100	−12.30928	0.03810	0.00492	−0.03453	1.49828	−17.23621
1.95	0.65762	−7.36744	0.06743	−0.02979	−0.02002	1.09309	−12.30752	0.03809	0.00492	−0.03451	1.50107	−17.23375
2.00	0.65879	−7.36655	0.06742	−0.02978	−0.02001	1.09483	−12.30606	0.03808	0.00492	−0.03450	1.50339	−17.23171
2.05	0.65976	−7.36582	0.06740	−0.02978	−0.02000	1.09628	−12.30484	0.03807	0.00491	−0.03448	1.50532	−17.23001
2.10	0.66057	−7.36520	0.06739	−0.02977	−0.01999	1.09748	−12.30383	0.03807	0.00491	−0.03447	1.50692	−17.22860
2.15	0.66124	−7.36469	0.06739	−0.02977	−0.01998	1.09848	−12.30298	0.03806	0.00491	−0.03446	1.50825	−17.22742
2.20	0.66180	−7.36427	0.06738	−0.02977	−0.01998	1.09932	−12.30228	0.03806	0.00491	−0.03445	1.50936	−17.22645
2.25	0.66226	−7.36392	0.06738	−0.02977	−0.01997	1.10001	−12.30170	0.03806	0.00491	−0.03445	1.51029	−17.22564
2.30	0.66265	−7.36362	0.06737	−0.02977	−0.01997	1.10059	−12.30122	0.03806	0.00491	−0.03444	1.51105	−17.22496
2.40	0.66324	−7.36317	0.06736	−0.02976	−0.01997	1.10146	−12.30048	0.03805	0.00491	−0.03443	1.51222	−17.22393
2.50	0.66365	−7.36287	0.06736	−0.02976	−0.01996	1.10207	−12.29997	0.03805	0.00491	−0.03443	1.51303	−17.22322
2.60	0.66393	−7.36265	0.06736	−0.02976	−0.01996	1.10249	−12.29961	0.03805	0.00491	−0.03442	1.51359	−17.22272
2.70	0.66412	−7.36250	0.06735	−0.02976	−0.01996	1.10278	−12.29937	0.03805	0.00491	−0.03442	1.51398	−17.22238
2.80	0.66426	−7.36240	0.06735	−0.02976	−0.01996	1.10298	−12.29920	0.03805	0.00491	−0.03442	1.51425	−17.22215
2.90	0.66435	−7.36233	0.06735	−0.02976	−0.01996	1.10312	−12.29908	0.03804	0.00491	−0.03442	1.51443	−17.22198
3.00	0.66442	−7.36228	0.06735	−0.02976	−0.01996	1.10322	−12.29900	0.03804	0.00491	−0.03442	1.51456	−17.22187
3.10	0.66446	−7.36225	0.06735	−0.02975	−0.01995	1.10329	−12.29895	0.03804	0.00491	−0.03442	1.51465	−17.22179

Table 8.16

h	$d_{24,2}^{(1)}$	$d_{24,2}^{(2)}$	$d_{24,2}^{(3)}$	$d_{24,2}^{(4)}$	$d_{24,3}^{(1)}$	$d_{24,3}^{(2)}$	$d_{24,3}^{(3)}$	$d_{24,3}^{(4)}$	$d_{24,4}^{(1)}$	$d_{24,4}^{(2)}$	$d_{24,4}^{(3)}$	$d_{24,4}^{(4)}$
1.05	0.14231	−0.02877	−0.00268	0.00150	−0.00598	0.07100	−0.02692	−0.00077	0.00167	−0.00349	0.04616	−0.02294
1.10	0.14341	−0.02882	−0.00268	0.00150	−0.00597	0.07100	−0.02692	−0.00077	0.00167	−0.00349	0.04616	−0.02294
1.15	0.14415	−0.02884	−0.00268	0.00150	−0.00597	0.07101	−0.02692	−0.00077	0.00167	−0.00349	0.04616	−0.02294
1.20	0.14465	−0.02886	−0.00268	0.00150	−0.00597	0.07101	−0.02692	−0.00077	0.00167	−0.00349	0.04616	−0.02294
1.25	0.14499	−0.02887	−0.00268	0.00150	−0.00597	0.07101	−0.02692	−0.00077	0.00167	−0.00349	0.04616	−0.02294
1.30	0.14522	−0.02887	−0.00268	0.00150	−0.00596	0.07101	−0.02692	−0.00077	0.00167	−0.00349	0.04616	−0.02294
1.35	0.14537	−0.02888	−0.00268	0.00150	−0.00596	0.07101	−0.02692	−0.00077	0.00167	−0.00349	0.04616	−0.02294
1.40	0.14548	−0.02888	−0.00268	0.00150	−0.00596	0.07101	−0.02692	−0.00077	0.00167	−0.00349	0.04616	−0.02294
1.45	0.14555	−0.02888	−0.00268	0.00150	−0.00596	0.07101	−0.02692	−0.00077	0.00167	−0.00349	0.04616	−0.02294
1.50	0.14560	−0.02888	−0.00268	0.00150	−0.00596	0.07101	−0.02692	−0.00077	0.00167	−0.00349	0.04616	−0.02294
1.55	0.14564	−0.02888	−0.00268	0.00150	−0.00596	0.07101	−0.02692	−0.00077	0.00167	−0.00349	0.04616	−0.02294
1.60	0.14566	−0.02888	−0.00268	0.00150	−0.00596	0.07101	−0.02692	−0.00077	0.00167	−0.00349	0.04616	−0.02294
1.65	0.14567	−0.02888	−0.00268	0.00150	−0.00596	0.07101	−0.02692	−0.00077	0.00167	−0.00349	0.04616	−0.02294
1.70	0.14568	−0.02888	−0.00268	0.00150	−0.00596	0.07101	−0.02692	−0.00077	0.00167	−0.00349	0.04616	−0.02294
1.75	0.14569	−0.02888	−0.00268	0.00150	−0.00596	0.07101	−0.02692	−0.00077	0.00167	−0.00349	0.04616	−0.02294
1.80	0.14570	−0.02888	−0.00268	0.00150	−0.00596	0.07101	−0.02692	−0.00077	0.00167	−0.00349	0.04616	−0.02294
1.85	0.14570	−0.02888	−0.00268	0.00150	−0.00596	0.07101	−0.02692	−0.00077	0.00167	−0.00349	0.04616	−0.02294
1.90	0.14570	−0.02888	−0.00268	0.00150	−0.00596	0.07101	−0.02692	−0.00077	0.00167	−0.00349	0.04616	−0.02294
1.95	0.14570	−0.02888	−0.00268	0.00150	−0.00596	0.07101	−0.02692	−0.00077	0.00167	−0.00349	0.04616	−0.02294
2.00	0.14570	−0.02888	−0.00268	0.00150	−0.00596	0.07101	−0.02692	−0.00077	0.00167	−0.00349	0.04616	−0.02294
2.05	0.14570	−0.02888	−0.00268	0.00150	−0.00596	0.07101	−0.02692	−0.00077	0.00167	−0.00349	0.04616	−0.02294
2.10	0.14571	−0.02888	−0.00268	0.00150	−0.00596	0.07101	−0.02692	−0.00077	0.00167	−0.00349	0.04616	−0.02294
2.15	0.14571	−0.02888	−0.00268	0.00150	−0.00596	0.07101	−0.02692	−0.00077	0.00167	−0.00349	0.04616	−0.02294
2.20	0.14571	−0.02888	−0.00268	0.00150	−0.00596	0.07101	−0.02692	−0.00077	0.00167	−0.00349	0.04616	−0.02294
2.25	0.14571	−0.02888	−0.00268	0.00150	−0.00596	0.07101	−0.02692	−0.00077	0.00167	−0.00349	0.04616	−0.02294
2.30	0.14571	−0.02888	−0.00268	0.00150	−0.00596	0.07101	−0.02692	−0.00077	0.00167	−0.00349	0.04616	−0.02294
2.40	0.14571	−0.02888	−0.00268	0.00150	−0.00596	0.07101	−0.02692	−0.00077	0.00167	−0.00349	0.04616	−0.02294
2.50	0.14571	−0.02888	−0.00268	0.00150	−0.00596	0.07101	−0.02692	−0.00077	0.00167	−0.00349	0.04616	−0.02294
2.60	0.14571	−0.02888	−0.00268	0.00150	−0.00596	0.07101	−0.02692	−0.00077	0.00167	−0.00349	0.04616	−0.02294
2.70	0.14571	−0.02888	−0.00268	0.00150	−0.00596	0.07101	−0.02692	−0.00077	0.00167	−0.00349	0.04616	−0.02294
2.80	0.14571	−0.02888	−0.00268	0.00150	−0.00596	0.07101	−0.02692	−0.00077	0.00167	−0.00349	0.04616	−0.02294
2.90	0.14571	−0.02888	−0.00268	0.00150	−0.00596	0.07101	−0.02692	−0.00077	0.00167	−0.00349	0.04616	−0.02294
3.00	0.14571	−0.02888	−0.00268	0.00150	−0.00596	0.07101	−0.02692	−0.00077	0.00167	−0.00349	0.04616	−0.02294
3.10	0.14571	−0.02888	−0.00268	0.00150	−0.00596	0.07101	−0.02692	−0.00077	0.00167	−0.00349	0.04616	−0.02294

Table 8.17

h	$d^{(1)}_{25,2}$	$d^{(2)}_{25,2}$	$d^{(3)}_{25,2}$	$d^{(4)}_{25,2}$	$d^{(1)}_{25,3}$	$d^{(2)}_{25,3}$	$d^{(3)}_{25,3}$	$d^{(4)}_{25,3}$	$d^{(1)}_{25,4}$	$d^{(2)}_{25,4}$	$d^{(3)}_{25,4}$	$d^{(4)}_{25,4}$
1.05	−8.60155	−7.44575	0.08408	−0.03729	−0.13281	−13.46622	−12.22041	0.05019	0.03197	−0.17664	−18.24156	−16.96145
1.10	−8.53833	−7.39379	0.08347	−0.03702	−0.13189	−13.37122	−12.13484	0.04984	0.03175	−0.17543	−18.11322	−16.84258
1.15	−8.48636	−7.35082	0.08296	−0.03679	−0.13114	−13.29270	−12.06412	0.04954	0.03157	−0.17443	−18.00716	−16.74434
1.20	−8.44356	−7.31527	0.08255	−0.03661	−0.13052	−13.22775	−12.00562	0.04930	0.03143	−0.17359	−17.91942	−16.66307
1.25	−8.40824	−7.28582	0.08220	−0.03645	−0.13001	−13.17397	−11.95717	0.04909	0.03130	−0.17291	−17.84676	−16.59577
1.30	−8.37906	−7.26140	0.08192	−0.03632	−0.12959	−13.12939	−11.91702	0.04893	0.03120	−0.17234	−17.78654	−16.54000
1.35	−8.35492	−7.24116	0.08168	−0.03622	−0.12924	−13.09243	−11.88373	0.04879	0.03112	−0.17186	−17.73661	−16.49375
1.40	−8.33493	−7.22436	0.08148	−0.03613	−0.12895	−13.06176	−11.85611	0.04867	0.03105	−0.17147	−17.69518	−16.45538
1.45	−8.31837	−7.21041	0.08132	−0.03606	−0.12871	−13.03631	−11.83318	0.04857	0.03099	−0.17115	−17.66079	−16.42353
1.50	−8.30464	−7.19883	0.08119	−0.03600	−0.12851	−13.01517	−11.81414	0.04850	0.03095	−0.17088	−17.63224	−16.39708
1.55	−8.29324	−7.18922	0.08107	−0.03595	−0.12835	−12.99761	−11.79833	0.04843	0.03091	−0.17065	−17.60852	−16.37511
1.60	−8.28378	−7.18122	0.08098	−0.03590	−0.12821	−12.98303	−11.78519	0.04837	0.03087	−0.17046	−17.58881	−16.35686
1.65	−8.27593	−7.17458	0.08090	−0.03587	−0.12810	−12.97091	−11.77427	0.04833	0.03085	−0.17031	−17.57244	−16.34169
1.70	−8.26941	−7.16906	0.08084	−0.03584	−0.12800	−12.96083	−11.76520	0.04829	0.03082	−0.17018	−17.55883	−16.32909
1.75	−8.26399	−7.16448	0.08078	−0.03582	−0.12792	−12.95246	−11.75765	0.04826	0.03081	−0.17007	−17.54752	−16.31861
1.80	−8.25948	−7.16066	0.08074	−0.03580	−0.12786	−12.94550	−11.75138	0.04823	0.03079	−0.16998	−17.53811	−16.30990
1.85	−8.25574	−7.15749	0.08070	−0.03578	−0.12780	−12.93971	−11.74617	0.04821	0.03078	−0.16991	−17.53029	−16.30266
1.90	−8.25263	−7.15485	0.08067	−0.03577	−0.12776	−12.93489	−11.74183	0.04819	0.03077	−0.16985	−17.52379	−16.29663
1.95	−8.25004	−7.15266	0.08065	−0.03575	−0.12772	−12.93089	−11.73823	0.04818	0.03076	−0.16980	−17.51839	−16.29163
2.00	−8.24789	−7.15084	0.08062	−0.03574	−0.12769	−12.92756	−11.73523	0.04817	0.03075	−0.16975	−17.51389	−16.28746
2.05	−8.24610	−7.14932	0.08061	−0.03574	−0.12766	−12.92480	−11.73274	0.04815	0.03074	−0.16972	−17.51015	−16.28400
2.10	−8.24461	−7.14806	0.08059	−0.03573	−0.12764	−12.92249	−11.73066	0.04815	0.03074	−0.16969	−17.50704	−16.28112
2.15	−8.24338	−7.14701	0.08058	−0.03572	−0.12762	−12.92058	−11.72894	0.04814	0.03073	−0.16967	−17.50445	−16.27872
2.20	−8.24235	−7.14614	0.08057	−0.03572	−0.12761	−12.91899	−11.72751	0.04813	0.03073	−0.16965	−17.50230	−16.27673
2.25	−8.24149	−7.14541	0.08056	−0.03572	−0.12760	−12.91766	−11.72631	0.04813	0.03073	−0.16963	−17.50051	−16.27507
2.30	−8.24078	−7.14481	0.08055	−0.03571	−0.12759	−12.91656	−11.72532	0.04812	0.03073	−0.16961	−17.49903	−16.27369
2.40	−8.23970	−7.14389	0.08054	−0.03571	−0.12757	−12.91488	−11.72381	0.04812	0.03072	−0.16959	−17.49676	−16.27159
2.50	−8.23895	−7.14325	0.08054	−0.03571	−0.12756	−12.91372	−11.72276	0.04811	0.03072	−0.16958	−17.49519	−16.27014
2.60	−8.23843	−7.14281	0.08053	−0.03570	−0.12755	−12.91292	−11.72204	0.04811	0.03072	−0.16957	−17.49410	−16.26914
2.70	−8.23807	−7.14251	0.08053	−0.03570	−0.12755	−12.91236	−11.72154	0.04811	0.03072	−0.16956	−17.49335	−16.26844
2.80	−8.23782	−7.14230	0.08052	−0.03570	−0.12754	−12.91198	−11.72119	0.04811	0.03072	−0.16956	−17.49283	−16.26796
2.90	−8.23765	−7.14215	0.08052	−0.03570	−0.12754	−12.91171	−11.72095	0.04811	0.03071	−0.16955	−17.49247	−16.26763
3.00	−8.23753	−7.14205	0.08052	−0.03570	−0.12754	−12.91153	−11.72079	0.04810	0.03071	−0.16955	−17.49222	−16.26740
3.10	−8.23745	−7.14198	0.08052	−0.03570	−0.12754	−12.91140	−11.72067	0.04810	0.03071	−0.16955	−17.49205	−16.26724

Tables of hydrodynamic coefficients 139

Table 8.18

h	μ_1	μ_2	μ_3	J_0	$\mu_4^{(1)}$	$\mu_4^{(2)}$	$\mu_4^{(3)}$	$\mu_4^{(4)}$	$\mu_5^{(1)}$	$\mu_5^{(2)}$	$\mu_5^{(3)}$	$\mu_5^{(4)}$
1.05	3.93354	2.23677	1.69678	0.38488	1.84798	−0.10429	0.03142	−0.01428	−0.32523	0.08828	−0.03703	0.02081
1.10	3.98109	2.21878	1.76231	0.40499	1.86171	−0.10223	0.03063	−0.01387	−0.36297	0.08789	−0.03686	0.02075
1.15	4.02719	2.20248	1.82470	0.42689	1.87479	−0.10029	0.02990	−0.01349	−0.39832	0.08747	−0.03671	0.02069
1.20	4.07156	2.18765	1.88392	0.45061	1.88719	−0.09849	0.02921	−0.01314	−0.43136	0.08705	−0.03657	0.02063
1.25	4.11402	2.17409	1.93993	0.47619	1.89892	−0.09680	0.02858	−0.01281	−0.46217	0.08662	−0.03644	0.02058
1.30	4.15442	2.16166	1.99277	0.50366	1.90998	−0.09524	0.02799	−0.01250	−0.49082	0.08621	−0.03632	0.02053
1.35	4.19271	2.15023	2.04247	0.53306	1.92038	−0.09378	0.02745	−0.01221	−0.51742	0.08580	−0.03621	0.02048
1.40	4.22884	2.13972	2.08912	0.56439	1.93014	−0.09243	0.02695	−0.01194	−0.54208	0.08541	−0.03610	0.02044
1.45	4.26282	2.13004	2.13279	0.59769	1.93926	−0.09118	0.02648	−0.01170	−0.56491	0.08505	−0.03600	0.02040
1.50	4.29469	2.12111	2.17358	0.63296	1.94778	−0.09002	0.02605	−0.01147	−0.58600	0.08470	−0.03592	0.02037
1.55	4.32449	2.11287	2.21162	0.67021	1.95572	−0.08895	0.02565	−0.01126	−0.60547	0.08437	−0.03583	0.02033
1.60	4.35229	2.10528	2.24702	0.70946	1.96311	−0.08797	0.02528	−0.01106	−0.62342	0.08407	−0.03576	0.02030
1.65	4.37817	2.09827	2.27990	0.75071	1.96996	−0.08706	0.02494	−0.01088	−0.63996	0.08379	−0.03569	0.02028
1.70	4.40223	2.09181	2.31041	0.79396	1.97632	−0.08622	0.02463	−0.01071	−0.65517	0.08352	−0.03562	0.02025
1.75	4.42453	2.08586	2.33868	0.83921	1.98221	−0.08544	0.02434	−0.01056	−0.66917	0.08328	−0.03556	0.02023
1.80	4.44520	2.08037	2.36482	0.88647	1.98765	−0.08473	0.02407	−0.01041	−0.68203	0.08305	−0.03551	0.02021
1.85	4.46430	2.07532	2.38898	0.93573	1.99268	−0.08408	0.02383	−0.01028	−0.69383	0.08284	−0.03546	0.02019
1.90	4.48195	2.07068	2.41128	0.98698	1.99732	−0.08348	0.02360	−0.01016	−0.70467	0.08265	−0.03541	0.02017
1.95	4.49824	2.06640	2.43184	1.04023	2.00159	−0.08293	0.02340	−0.01005	−0.71460	0.08247	−0.03537	0.02015
2.00	4.51325	2.06247	2.45078	1.09546	2.00553	−0.08243	0.02321	−0.00995	−0.72371	0.08231	−0.03533	0.02014
2.05	4.52707	2.05886	2.46821	1.15268	2.00915	−0.08196	0.02303	−0.00986	−0.73206	0.08216	−0.03530	0.02012
2.10	4.53978	2.05554	2.48425	1.21187	2.01248	−0.08154	0.02287	−0.00977	−0.73971	0.08203	−0.03527	0.02011
2.15	4.55147	2.05249	2.49898	1.27303	2.01554	−0.08115	0.02273	−0.00969	−0.74671	0.08190	−0.03524	0.02010
2.20	4.56221	2.04970	2.51251	1.33615	2.01835	−0.08079	0.02259	−0.00962	−0.75311	0.08178	−0.03521	0.02009
2.25	4.57207	2.04713	2.52493	1.40123	2.02092	−0.08047	0.02247	−0.00955	−0.75898	0.08168	−0.03518	0.02008
2.30	4.58111	2.04478	2.53633	1.46824	2.02329	−0.08017	0.02236	−0.00949	−0.76434	0.08158	−0.03516	0.02007
2.40	4.59701	2.04065	2.55635	1.60808	2.02744	−0.07964	0.02216	−0.00938	−0.77373	0.08141	−0.03512	0.02005
2.50	4.61035	2.03719	2.57316	1.75559	2.03092	−0.07921	0.02200	−0.00930	−0.78157	0.08127	−0.03509	0.02004
2.60	4.62154	2.03429	2.58724	1.91071	2.03384	−0.07884	0.02186	−0.00922	−0.78812	0.08115	−0.03506	0.02003
2.70	4.63090	2.03186	2.59903	2.07339	2.03628	−0.07854	0.02174	−0.00916	−0.79358	0.08105	−0.03504	0.02002
2.80	4.63873	2.02984	2.60889	2.24355	2.03832	−0.07828	0.02165	−0.00911	−0.79814	0.08097	−0.03502	0.02001
2.90	4.64527	2.02814	2.61712	2.42115	2.04003	−0.07807	0.02157	−0.00906	−0.80194	0.08090	−0.03500	0.02000
3.00	4.65073	2.02673	2.62400	2.60613	2.04145	−0.07789	0.02150	−0.00903	−0.80510	0.08085	−0.03499	0.02000
3.10	4.65528	2.02555	2.62973	2.79844	2.04264	−0.07774	0.02144	−0.00900	−0.80774	0.08080	−0.03498	0.01999

Table 8.19

h	$\mu_6^{(1)}$	$\mu_6^{(2)}$	$\mu_6^{(3)}$	$\mu_6^{(4)}$	$\mu_7^{(1)}$	$\mu_7^{(2)}$	$\mu_7^{(3)}$	$\mu_7^{(4)}$	$\tilde{P}_1$	$\tilde{P}_2$	$\tilde{P}_3$	$\tilde{P}_4$
1.05	0.64796	−0.19503	0.08503	−0.04746	−2.42352	0.43905	−0.17654	0.09496	0.20164	−0.00943	0.00229	−0.00088
1.10	0.68514	−0.19651	0.08533	−0.04758	−2.45637	0.44128	−0.17744	0.09549	0.20694	−0.00945	0.00229	−0.00088
1.15	0.71978	−0.19782	0.08561	−0.04770	−2.48759	0.44344	−0.17833	0.09602	0.21188	−0.00946	0.00229	−0.00088
1.20	0.75199	−0.19898	0.08587	−0.04781	−2.51717	0.44552	−0.17920	0.09653	0.21647	−0.00947	0.00229	−0.00088
1.25	0.78190	−0.20002	0.08611	−0.04790	−2.54510	0.44753	−0.18004	0.09702	0.22074	−0.00948	0.00229	−0.00088
1.30	0.80962	−0.20096	0.08633	−0.04800	−2.57140	0.44944	−0.18086	0.09749	0.22470	−0.00948	0.00229	−0.00088
1.35	0.83529	−0.20181	0.08654	−0.04808	−2.59610	0.45126	−0.18164	0.09794	0.22836	−0.00949	0.00229	−0.00088
1.40	0.85903	−0.20257	0.08673	−0.04816	−2.61925	0.45299	−0.18238	0.09837	0.23175	−0.00949	0.00229	−0.00088
1.45	0.88096	−0.20327	0.08690	−0.04823	−2.64089	0.45463	−0.18308	0.09878	0.23489	−0.00949	0.00229	−0.00088
1.50	0.90119	−0.20390	0.08706	−0.04830	−2.66107	0.45617	−0.18375	0.09916	0.23778	−0.00949	0.00229	−0.00088
1.55	0.91984	−0.20447	0.08721	−0.04836	−2.67987	0.45762	−0.18438	0.09952	0.24045	−0.00950	0.00229	−0.00088
1.60	0.93701	−0.20499	0.08734	−0.04842	−2.69734	0.45898	−0.18497	0.09985	0.24290	−0.00950	0.00229	−0.00088
1.65	0.95282	−0.20547	0.08747	−0.04847	−2.71356	0.46025	−0.18552	0.10017	0.24517	−0.00950	0.00229	−0.00088
1.70	0.96735	−0.20591	0.08758	−0.04852	−2.72859	0.46144	−0.18603	0.10046	0.24725	−0.00950	0.00229	−0.00088
1.75	0.98071	−0.20630	0.08769	−0.04856	−2.74249	0.46254	−0.18651	0.10074	0.24916	−0.00950	0.00229	−0.00088
1.80	0.99298	−0.20667	0.08779	−0.04860	−2.75535	0.46357	−0.18696	0.10099	0.25091	−0.00950	0.00229	−0.00088
1.85	1.00424	−0.20700	0.08788	−0.04864	−2.76722	0.46453	−0.18737	0.10122	0.25253	−0.00950	0.00229	−0.00088
1.90	1.01457	−0.20730	0.08796	−0.04867	−2.77816	0.46541	−0.18775	0.10144	0.25401	−0.00950	0.00229	−0.00088
1.95	1.02404	−0.20758	0.08803	−0.04871	−2.78825	0.46623	−0.18811	0.10164	0.25536	−0.00950	0.00229	−0.00088
2.00	1.03272	−0.20784	0.08810	−0.04873	−2.79753	0.46698	−0.18844	0.10183	0.25660	−0.00950	0.00229	−0.00088
2.05	1.04067	−0.20807	0.08816	−0.04876	−2.80607	0.46768	−0.18874	0.10200	0.25774	−0.00950	0.00229	−0.00088
2.10	1.04795	−0.20828	0.08822	−0.04879	−2.81392	0.46833	−0.18902	0.10216	0.25879	−0.00950	0.00229	−0.00088
2.15	1.05462	−0.20847	0.08827	−0.04881	−2.82113	0.46892	−0.18927	0.10230	0.25974	−0.00950	0.00229	−0.00088
2.20	1.06072	−0.20865	0.08832	−0.04883	−2.82775	0.46946	−0.18951	0.10244	0.26062	−0.00950	0.00229	−0.00088
2.25	1.06630	−0.20881	0.08837	−0.04885	−2.83382	0.46997	−0.18973	0.10256	0.26142	−0.00950	0.00229	−0.00088
2.30	1.07141	−0.20896	0.08841	−0.04886	−2.83939	0.47043	−0.18993	0.10267	0.26215	−0.00950	0.00229	−0.00088
2.40	1.08034	−0.20922	0.08848	−0.04889	−2.84917	0.47124	−0.19028	0.10287	0.26343	−0.00950	0.00229	−0.00088
2.50	1.08781	−0.20944	0.08854	−0.04892	−2.85738	0.47192	−0.19057	0.10304	0.26450	−0.00950	0.00229	−0.00088
2.60	1.09404	−0.20962	0.08859	−0.04894	−2.86425	0.47250	−0.19082	0.10318	0.26539	−0.00950	0.00229	−0.00088
2.70	1.09925	−0.20977	0.08863	−0.04896	−2.87000	0.47298	−0.19103	0.10330	0.26613	−0.00950	0.00229	−0.00088
2.80	1.10358	−0.20989	0.08866	−0.04897	−2.87480	0.47338	−0.19120	0.10339	0.26676	−0.00950	0.00229	−0.00088
2.90	1.10720	−0.21000	0.08869	−0.04899	−2.87881	0.47372	−0.19135	0.10348	0.26727	−0.00950	0.00229	−0.00088
3.00	1.11021	−0.21008	0.08871	−0.04899	−2.88216	0.47400	−0.19147	0.10355	0.26771	−0.00950	0.00229	−0.00088
3.10	1.11272	−0.21016	0.08873	−0.04900	−2.88496	0.47424	−0.19157	0.10360	0.26807	−0.00950	0.00229	−0.00088

References

1. K.A. Abgaryan, I.M. Rappoport. *Rockets dynamics*. Moscow: Mashinostroenie, 1969 (in Russian).
2. H.N. Abramson. *The dynamics behavior of liquid in moving containers with applications to space vehicle technology:* Tech. Rep. NASA, SP-106. NASA, Washington, D.C., 1966.
3. H.N. Abramson. Dynamics of contained lliquid: A personal odyssey. *Applied Mechanics Reviews*, 56(1):R1–R7, 2003.
4. H.N. Abramson, W.H. Chu, and D.D. Kana. Some studies of nonlinear lateral sloshing in rigid containers. *Journal of Applied Mechanics*, 33(4):66–74, 1966.
5. M. Antuono, B. Bouscasse, A. Colagrossi, and C. Lugni. Two-dimensional modal method for shallow-water sloshing in rectangular basins. *Journal of Fluid Mechanics*, 700:419–440, 2012.
6. M. Arai. Experimental and numerical studies of sloshing in liquid cargo tanks with internal structures. *IHI Engineering Review*, 19:1–9, 1986.
7. K. Barkowiak, B. Gampert, and J. Siekmann. On liquid motion in a circular cylinder with horizontal axis. *Acta Mechanica*, 54:207–220, 1985.
8. M. Barnyak, I. Gavrilyuk, M. Hermann, and A. Timokha. Analytical velocity potentials in cells with a rigid spherical wall. *Zeitschrift für Angewandte Mathematik und Mechanik*, 91(1):38–45, 2011.
9. M.Ya. Barnyak and O.P. Leshchuk. Construction of solutions of the problem of free oscillations of viscous fluid in a half-filled spherical tank. *Nonlinear Oscillations*, 11(4):461–483, 2008.
10. H. Bateman. *Partial differential equations of mathematical physics*. Cambridge University Press, 1932.
11. H. Bateman. *Partial differential equations*. New York: Dover Publications, 1944.
12. H.F. Bauer. Flüssigkeitsschwingungen in Kegelbehälterformen. *Acta Mechanica*, 43:185–200, 1982.
13. H.F. Bauer and W. Eidel. Non-linear liquid motion in conical container. *Acta Mechanica*, 464(73, 1–4):11–31, 1988.
14. R.R. Berlot. Production of rotation in confined liquid through translational motion of the boundaries. *Jounal of Applied Mechanics*, 26:513–516, 1959.
15. J. Billingham. Nonlinear sloshing in zero gravity. *Journal of Fluid Mechanics*, 464:365–391, 2002.
16. J. Billingham. On a model for the motion of a contact line on a smooth solid surface. *European Journal of Applied Mathematics*, 17:347–382, 2006.
17. R.D. Blevins. *Applied fluid dynamics*. Krieger Publishing Company, Malabar, FL, 1992.
18. G.I. Bogomaz and S.A. Sirota. *Oscillations of a liquid in containers: methods and results of experimental studies*. Dnepropetrovsk: National Space Agency of Ukraine, 2002 (in Russian).
19. I. Bogoryad. *Oscillations of a viscous liquid in a cavity of a rigid body*. Tomsk: Tomsk University, 1999 (in Russian).

20. I. Bogoryad, I. Druzhinin, and S. Chakhlov. *Study of transient processes with large disturbances of the free surface of a liquid in a closed compartment.* Moscow: Mashinostroenie, 194–203, (in Russian), 1986.
21. I.V. Bogoryad, I.A. Druzhinin, G.Z. Druzhinina, and E.E. Libin. *Introduction to the dynamics of vessels with a liquid.* Tomsk: Tomsk University, 1977 (in Russian).
22. J. Bouvard, W. Herreman, and F. Moisy. Mean mass transport in an orbitally shaken cylindrical container. *Physical Review Fluids*, 2 (Paper No 084801), 2017.
23. T.J. Bridges. On secondary bifurcation of three-dimensional standing waves. *SIAM Journal of Applied Mathematics*, 47:40–59, 1986.
24. T.J. Bridges. Secondary bifurcation and change of type for type three dimensional standing waves in finite depth. *Journal of Fluid Mechanics*, 179:137–153, 1987.
25. P.J. Bryant. Nonlinear progressive waves in a circular basin. *Journal of Fluid Mechanics*, 205:453–467, 1989.
26. B. Budiansky. Sloshing of liquid in circular canals and spherical tanks. *Journal of Aerospace Sciences*, 27(3):161–172, 1960.
27. E. Buldakov. Lagrangian modelling of fluid sloshing in moving tanks. *Journal of Fluids and Structures*, 45:1–14, 2014.
28. A. Cariou and G. Casella. Liquid sloshing in ship tanks: a comparative study of numerical simulation. *Marine Structures*, 12(3):183–198, 1999.
29. F. Casciati, A.D. Stefano, and E. Matta. Simulating a conical tuned liquid damper. *Simulation Modelling Practice and Theory*, 11:353–370, 2003.
30. K.M. Case and W.C Parkinson. Damping of surface waves in an incompressible liquid. *Journal of Fluid Mechanics*, 2:172–184, 1957.
31. F. Chernous'ko. *Motion of a rigid body with cavities containing viscous liquid.* Moscow: Nauka, 1968 (in Russian).
32. F.L. Chernous'ko. On free oscillations of a viscous fluid in a vessel. *Journal of Applied Mathematics and Mechanics*, 30(5):990–1003, 1966.
33. M. Chernova and A. Timokha. Sloshing in a two-dimensional circular tank. Weakly-nonlinear modal equations. *Transactions of Institute of Mathematics of NAS of Ukraine*, 10(3):262–283, 2013.
34. M.O. Chernova, I.A. Lukovsky, and A.N. Timokha. Generalizing the multimodal method for the levitating drop dynamics. *ISRN Mathematical Physics*, Article ID 869070:1–19, 2012.
35. W. Chester. Resonant oscillation of water waves I. Theory. *Proceeding of Royal Society, London*, 308:5–22, 1968.
36. W. Chester and J.A. Bones. Resonant oscillation of water waves. II. Experiment. *Proceeding of Royal Society, London*, 306:23–30, 1968.
37. R.M. Cooper. Dynamics of liquid in moving containers. *ARS Journal*, 30(8):725–729, 1960.
38. A.D. Craik. The origin of water wave theory. *Annual Review of Fluid Mechanics*, 36:1–28, 2004.
39. L.V. Dokuchaev. On the solution of a boundary value problem on the sloshing of a liquid in conical cavities. *Applied Mathematics and Mechanics (PMM)*, 28(1):601–602, 1964.

40. I.A. Druzhinin and S.V. Chakhlov. Computation of nonlinear sloshing (Cauchy problem). In *Dynamics of elastic and rigid bodies interacting with a liquid / Proceedings of IV Seminar, September, 15–18, 1980 Tomsk: 1981*, 85–91, 1981 (in Russian).
41. A. Ducci and W.H. Weheliye. Orbitally shaken bioreactors – viscosity effects on flow characteristics. *AIChE Journal*, 60:3951–3968, 2014.
42. M. Eastham. An eigenvalue problem with parameter in the boundary condition. *Quarterly Journal of Mathematics*, 13:304–320, 1962.
43. C.R. Easton and I. Catton. Initial value technique in free-surface hydrodynamics. *Journal of Computational Physics*, 9:424–439, 1972.
44. T. Eguchi and O Niho. A numerical simulation of 2-dimensional sloshing problem. *Mitsui Zosen Technical Review*, N 138:1–19, 1989.
45. R. Elahi, M. Passandideh-Fard, and A. Javanshir. Simulation of liquid sloshing in 2D containers using the volume of fluid method. *Ocean Engineering*, 96:226–244, 2015.
46. A. Faller. The constant-V vortex. *Journal of Ship Research*, 434:167–180, 2001.
47. O.M. Faltinsen. A nonlinear theory of sloshing in rectangular tanks. *Journal of Ship Research*, 18(4):224–241, 1974.
48. O.M. Faltinsen. A nonlinear method of sloshing in tanks with two-dimensional flow. *Journal of Ship Research*, 22(3):193–202, 1978.
49. O.M. Faltinsen, R. Firoozkoohi, and A.N. Timokha. Analytical modeling of liquid sloshing in a two-dimensional rectangular tank with a slat screen. *Journal of Engineering Mathematics*, 70(1-2):93–109, 2011.
50. O.M. Faltinsen, R. Firoozkoohi, and A.N. Timokha. Effect of central slotted screen with a high solidity ratio on the secondary resonance phenomenon for liquid sloshing in a rectangular tank. *Physics of Fluids*, 23:Art. No. 062106, 2011.
51. O.M. Faltinsen, R. Firoozkoohi, and A.N. Timokha. Steady-state liquid sloshing in a rectangular tank with a slat-type screen in the middle: quasi-linear modal analysis and experiments. *Physics of Fluids*, 23:Art. No. 042101, 2011.
52. O.M. Faltinsen, I.A. Lukovsky, and A.N. Timokha. Resonant sloshing in an upright tank. *Journal of Fluid Mechanics*, 804:608–645, 2016.
53. O.M. Faltinsen and O.F. Rognebakke. Sloshing. In *NAV 2000: International Conference on Ship and Shipping Research: 13th Congress, 19-22 September 2000, Venice (Italy): Proceedings, 2000*, 2000.
54. O.M. Faltinsen, O.F. Rognebakke, I.A. Lukovsky, and A.N. Timokha. Multidimensional modal analysis of nonlinear sloshing in a rectangular tank with finite water depth. *Journal of Fluid Mechanics*, 407:201–234, 2000.
55. O.M. Faltinsen, O.F. Rognebakke, and A.N. Timokha. Resonant three-dimensional nonlinear sloshing in a square base basin. *Journal of Fluid Mechanics*, 487:1–42, 2003.
56. O.M. Faltinsen, O.F. Rognebakke, and A.N. Timokha. Classification of three-dimensional nonlinear sloshing in a square-base tank with finite depth. *Journal of Fluids and Structures*, 20(1):81–103, 2005.
57. O.M. Faltinsen, O.F. Rognebakke, and A.N. Timokha. Resonant three-dimensional nonlinear sloshing in a square base basin. Part 2. Effect of higher modes. *Journal of Fluid Mechanics*, 523:199–218, 2005.

58. O.M. Faltinsen, O.F. Rognebakke, and A.N. Timokha. Resonant three-dimensional nonlinear sloshing in a square base basin. Part 3. Base ratio perturbation. *Journal of Fluid Mechanics*, 551:93–116, 2006.
59. O.M. Faltinsen, O.F. Rognebakke, and A.N. Timokha. Transient and steady-state amplitudes of resonant three-dimensional sloshing in a square base tank with a finite fluid depth. *Physics of Fluids*, 18:Art. No. 012103, 2006.
60. O.M. Faltinsen and A.N. Timokha. Adaptive multimodal approach to nonlinear sloshing in a rectangular tank. *Journal of Fluid Mechanics*, 432:167–200, 2001.
61. O.M. Faltinsen and A.N. Timokha. Asymptotic modal approximation of nonlinear resonant sloshing in a rectangular tank with small fluid depth. *Journal of Fluid Mechanics*, 470:319–357, 2002.
62. O.M. Faltinsen and A.N. Timokha. *Sloshing*. Cambridge University Press, 2009.
63. O.M. Faltinsen and A.N. Timokha. A multimodal method for liquid sloshing in a two-dimensional circular tank. *Journal of Fluid Mechanics*, 665:457–479, 2010.
64. O.M. Faltinsen and A.N. Timokha. Natural sloshing frequencies and modes in a rectangular tank with a slat-type screen. *Journal of Sound and Vibration*, 330:1490–1503, 2011.
65. O.M. Faltinsen and A.N. Timokha. Analytically approximate natural sloshing modes for a spherical tank shape. *Journal of Fluid Mechanics*, 703:391–401, 2012.
66. O.M. Faltinsen and A.N. Timokha. On sloshing modes in a circular tank. *Journal of Fluid Mechanics*, 695:467–477, 2012.
67. O.M. Faltinsen and A.N. Timokha. Multimodal analysis of weakly nonlinear sloshing in a spherical tank. *Journal of Fluid Mechanics*, 719:129–164, 2013.
68. O.M. Faltinsen and A.N. Timokha. Nonlinear sloshing in a spherical tank. In *OMAE 2013. Proceeding of the ASME 32nd International Conference on Ocean, Offshore and Arctic Engineering. June 9-14, Nantes, France*, 2013.
69. O.M. Faltinsen and A.N. Timokha. Analytically approximate natural sloshing modes and frequencies in two dimensional tanks. *European Journal of Mechanics B/Fluids*, 47:176–187, 2014.
70. O.M. Faltinsen and A.N. Timokha. Resonant three-dimensional Nonlinear sloshing in a square-base basin. Part 4. Oblique forcing and linear viscous damping. *Journal of Fluid Mechanics*, 822:139–169, 2017.
71. O.M. Faltinsen and A.N. Timokha. An inviscid analysis of the Prandtl azimuthal mass transport during swirl-type sloshing. *Journal of Fluid Mechanics*, 865:884–903, 2019.
72. S.F. Feschenko, I.A. Lukovsky, B.I. Rabinovich, and L.V. Dokuchaev. *Methods of determining the added liquid mass in mobile cavities*. Kiev: Naukova Dumka, 1969 (in Russian).
73. L.L. Fontenot. *Dynamic stability of space vehicles. Vol. 7. The dynamics of liquid in fixed and moving containers: Tech. Rep. NASA, CR-941*. Washington: NASA, 1968.
74. L.K. Forbes. Sloshing of an ideal fluid in a horizontally forced rectangular tank. *Journal of Engineering Mathematics*, 66(4):395–412, 2010.
75. D. Fultz. An experimental note on finite-amplitude standing gravity waves. *Journal of Fluid Mechanics*, 13(2):192–212, 1962.

76. M. Funakoshi and S. Inoue. Surface waves due to resonant oscillation. *Journal of Fluid Mechanics*, 192:219–247, 1988.
77. M. Funakoshi and S. Inoue. Bifurcations in resonantly forced water waves. *European Journal of Mechanics, B/Fluids*, 10:31–36, 1991.
78. I. Gavrilyuk, M. Hermann, and I. Lukovsky. Weakly-nonlinear sloshing in a truncated circular conical tank. *Fluid Dynamics Research*, 45:1–30, 2013.
79. I. Gavrilyuk, M. Hermann, I. Lukovsky, D. Ovchynnykov, and A. Timokha. Computer-based multimodal modeling of liquid sloshing in a circular cylindrical tank. *Reports on Numerical Mathematics, FSU Jena*, N 10-03:1–14, 2010.
80. I. Gavrilyuk, M. Hermann, I. Lukovsky, O. Solodun, and A.Timokha. Natural sloshing frequencies in rigid truncated conical tanks. *Engineering Computations*, 25(6):518–540, 2008.
81. I. Gavrilyuk, I. Lukovsky, and A.N.Timokha. A multimodal approach to nonlinear sloshing in a circular cylindrical tank. *Hybrid Methods in Engineering*, 2(4):463–483, 2000.
82. I. Gavrilyuk, I. Lukovsky, and A.N. Timokha. Sloshing in circular conical tank. *Hybrid Methods in Engineering*, 3(4):322–378, 2001.
83. I. Gavrilyuk, I. Lukovsky, Y. Trotsenko, and A.Timokha. Sloshing in a vertical circular cylindrical tank with an annular baffle. Part 1. Linear fundamental solutions. *Journal of Engineering Mathematics*, 54:71–88, 2006.
84. I. Gavrilyuk, I. Lukovsky, Y. Trotsenko, and A.Timokha. Sloshing in a vertical circular cylindrical tank with an annular baffle. Part 2. Nonlinear resonant waves. *Journal of Engineering Mathematics*, 57:57–78, 2007.
85. I. Gavrilyuk, M. Hermann I. Lukovsky, O. Solodin, and A. Timokha. Multimodal method for linear liquid sloshing in a rigid tapered conical tank. *Engineering Computations*, 29(2):198–220, 2012.
86. I.P. Gavrilyuk, I.A. Lukovsky, and A.N. Timokha. Linear and nonlinear sloshing in a circular conical tank. *Fluid Dynamics Research*, 37:399–429, 2005.
87. L. Guo and K. Morita. Numerical simulation of 3D sloshing in a liquid-solid mixture using particle methods. *International Journal for Numerical Methods in Engineering*, 95(9):771–790, 2013.
88. J.A. Hamelin, J.S. Love, M.J. Tait, and J.C. Wilson. Tuned liquid dampers with a Keulegen-Carpenter number-dependent screen drag coefficient. *Journal of Fluids and Structures*, 43:271–286, 2013.
89. R. Hargneaves. A pressure-integral as kinetic potential. *Philosophical Magazine*, 16:436–444, 1908.
90. D.M. Henderson and J.W. Miles. Faraday waves in 2:1 resonance. *Journal of Fluid Mechanics*, 222:449–470, 1991.
91. D.M. Henderson and J.W. Miles. Surface-wave damping in a circular cylinder with a fixed contact angle. *Journal of Fluid Mechanics*, 275:285–299, 1994.
92. M. Hermann and A.Timokha. Modal modelling of the nonlinear resonant sloshing in a rectangular tank I: A single-dominant model. *Mathematical Models and Methods in Applied Sciences*, 15(9):1431–1458, 2005.
93. M. Hermann and A.Timokha. Modal modelling of the nonlinear resonant fluid sloshing in a rectangular tank II: Secondary resonance. *Mathematical Models and Methods in Applied Sciences*, 18(11):1845–1867, 2008.

94. P. Holmes. Chaotic motions in a weakly nonlinear model for surface waves. *Journal of Fluid Mechanics*, 162:365–388, 1986.
95. R.E. Hutton. An investigation of nonlinear, nonplanar oscillations of fluid in cylindrical container: Tech. rep. NASA; D-1870. NASA, 1963.
96. R.E. Hutton. Fluid-particle during rotary sloshing. *Journal of Applied Mechanics, Transactions ASME*, 31:145–163, 1964.
97. T. Ikeda. Nonlinear parametric vibrations of an elastic structure with a rectangular liquid tank. *Nonlinear Dynamics*, 33(1):43–70, 2003.
98. T. Ikeda. Autoparametric resonances in elastic sructures carrying two rectangular tanks partially filled with liquid. *Journal of Sound and Vibration*, 302(4–5):657–682, 2007.
99. T. Ikeda and R.A. Ibrahim. Nonlinear random responses of a structure parametrically coupled with liquid sloshing in a cylindrical tank. *Journal of Sound and Vibration*, 284:75–102, 2005.
100. T. Ikeda, R.A. Ibrahim, Y. Harata, and T. Kuriyama. Nonlinear liquid sloshing in a square tank subjected to obliquely horizontal excitation. *Journal of Fluid Mechanics*, 700:304–328, 2012.
101. T. Ikeda and N. Nakagawa. Nonlinear vibrations of a structure caused by water sloshing in a rectangular tank. *Journal of Sound and Vibration*, 201(1):23–41, 1997.
102. N. Joukowski. On motions of a rigid body with cavity filled by homogeneous liquid. *Journal of Russian Physical-Mathematical Society*, **XVI**:30–85 (in Russian), 1885.
103. G. Keulegan. Energy dissipation in standing waves in rectangular basins. *Journal of Fluid Mechanics*, 6:33–50, 1959.
104. S.G. Klein. Oscillations of a viscous fluid in a container. *Doklady Akademii Nauk SSSR*, 159:262–265, 1964.
105. A. Kolaei, S. Rakheja, and M.J. Richard. Effects of tank cross-section on dynamic fluid slosh loads and roll stability of a partly-filled tank truck. *European Journal of Mechanics, B/Fluids*, 46:46–58, 2014.
106. A. Kolaei, S. Rakheja, and M.J. Richard. A coupled multimodal and boundary-element method for analysis of anti-slosh effectiveness of partial baffles in a partly-filled container. *Computers & Fluids*, 107:43–58, 2015.
107. K.S. Kolesnikov. *Propellant rocket as a control object*. Moscow: Mashinostroenie, 1969 (in Russian).
108. A. Komarenko. Asymptotic expansion of eigenfunctions of a problem with a parameter in the boundary conditions in a neighborhood of angular boundary points. *Ukrainian Mathematical Journal*, 32(5):433–437, 1980.
109. A. Komarenko. Asymptotics of solutions of sectral problems of hydrodynamics in the neighborhood of angular points. *Ukrainian Mathematical Journal*, 50(6):912–921, 1998.
110. N. Kopachevsky and S. Klein. *Operator approach to linear problems of hydrodynamics. Volume 2: Nonself-adjoint problems for viscous fluid*. Basel-Boston-Berlin: Birkhauser Verlag, 2003.
111. N.D. Kopachevsky and S.G. Klein. *Operator approach to linear problems of hydrodynamics. Volume 1: Self-adjoint problems for an ideal fluid*. Basel-Boston-Berlin: Birkhauser Verlag, 2003.

References

112. T.S. Krasnopolskaya and A.Y. Shvets. Dynamical chaos for a limited power supply for fluid oscillations in cylindrical tanks. *Journal of Sound and Vibration*, 322:532–553, 2009.
113. V.D. Kubenko and P.S. Koval'chuk. Nonlinear problems of the dynamics of elastic shells partially filled with a liquid. *International Applied Mechanics*, 36(4):421–448, 2000.
114. M. La Rocca, M. Scortino, and M.A. Boniforty. A fully nonlinear model for sloshing in a rotating container. *Fluid Dynamics Research*, 27:25–229, 2000.
115. L.D. Landau and E.M. Lifshitz. *Hydrodynamics*. Moscow: Nauka, 1986 (in Russian).
116. D.Y. Lee, H.S. Choi, and O.M. Faltinsen. A study on the sloshing effect on the motion of 2d boxes in regular waves. *Journal of Hydrodynamics*, 22(5):429–433, 2010.
117. D.Y. Lee, G.N. Jo, Y.H. Kim, H.S. Choi, and O.M. Faltinsen. The effect of sloshing on the sway motions of 2D rectangular cylinders. *Journal of Marine Science and Technology*, 16(3):323–330, 2011.
118. O.S. Limarchenko. Direct method for solution of nonlinear dynamic problem on the motion of a tank with fluid. *Dopovidi Akademii Nauk Ukrains'koi RSR, Series A*, 11(11):99–102, 1978.
119. O.S. Limarchenko. Variational-method investigation of problems of nonlinear dynamics of a reservoir with a liquid. *Soviet Applied Mechanics*, 1(16):74–79, 1980.
120. O.S. Limarchenko. Application of a variational method to the solution of nonlinear problems of the dynamics of combined motions of a tank with fluid. *Soviet Applied Mechanics*, 11(19):1021–1025, 1983.
121. O.S. Limarchenko. Nonlinear properties for dynamic behavior of liquid with a free surface in a rigid moving tank. *International Journal of Nonlinear Sciences and Numerical Simulation*, 1(2):105–118, 2000.
122. O.S. Limarchenko. Specific features of aplication of perturbation techniques in problems of nonlinear oscillations of a liquid with free surface in cavities of noncylindrical shape. *Ukrainian Mathematical Journal*, 59:45–69, 2007.
123. O.S. Limarchenko. Nonlinear wave generation on a fluid in a moving parabolic tank. *International Applied Mechanics*, 46(8):864–868, 2011.
124. O.S. Limarchenko and V.V. Yasinskii. *Nonlinear dynamics of constructions with a fluid*. Kiev: Kiev Polytechnical University, 1996 (in Russian).
125. J.S. Love and M.J. Tait. Nonlinear simulation of a tuned liquid damper with damping screens using a modal expansion technique. *Journal of Fluids and Structures*, 26(7–8):1058–1077, 2010.
126. J.S. Love and M.J. Tait. Non-linear multimodal model for tuned liquid dampers of arbitrary tank geometry. *International Journal of Non-Linear Mechanics*, 46(8):1065–1075, 2011.
127. J.S. Love and M.J. Tait. Nonlinear multimodal model for TLD of irregular tank geometry and small fluid depth. *Journal of Fluids and Structures*, 43:83–99, 2013.
128. J.S. Love and M.J. Tait. Parametric depth ratio study on tuned liquid dampers: Fluid modelling and experimenta work. *Computers & Fluids*, 79:13–26, 2013.

129. J.G. Luke. A variational principle for a fluid with a free surface. *Journal of Fluid Mechanics*, 27:395–397, 1967.
130. I.A. Lukovskii. Variational formulation of nonlinear dynamic boundary problems of a finite liquid volume performing specified spatial motion. *Soviet Applied Mechanics*, 16(2):164–169, 1980.
131. I.A. Lukovskii. Approximate method of solution of nonlinear problems in the dynamics of a liquid in a vessel executing a prescribed motion. *Soviet Applied Mechanics*, 17(2):172–178, 1981.
132. I.A. Lukovskii. Variational methods of solving dynamic problems for fluid-containing bodies. *International Applied Mechanics*, 40(10):1092–1128, 2004.
133. I. Lukovsky, D. Ovchynnykov, and A. Timokha. Asymptotic nonlinear multimodal method for liquid sloshing in an upright circular cylindrical tank. Part 1: Modal equations. *Nonlinear Oscillations*, 14(4):512–525, 2012.
134. I. Lukovsky and A. Pilkevich. *On liquid motions in an upright oscillating circular tank*. Numerical-analytical studies of the dynamics and stability of multidimensional systems. Kiev: Institute of Mathematics, 1985 (in Russian).
135. I. Lukovsky, O. Solodun, and A. Timokha. Eigen oscillations of a liquid sloshing in truncated conical tanks. *Acoustic Bulletin*, 9:42–61 (in Russian), 2006.
136. I. Lukovsky and A. Timokha. Modal modeling of nonlinear sloshing in tanks with non-vertical walls. Non-conformal maping technique. *International Journal of Fluid Mechanics Research*, 29(2):216–242, 2002.
137. I. Lukovsky and A. Timokha. Combining Narimanov–Moiseev' and Lukovsky–Miles' schemes for nonlinear liquid sloshing. *Journal of Numerical and Applied Mathematics*, 105:69–82, 2011.
138. I.A. Lukovsky. On constructing a solution of a nonlinear problem on free oscillations of a liquid in basins of arbirtay shape. *Dopovidi AN UkrSSR, Ser. A*, (3):207–210 (in Russian) 1969.
139. I.A. Lukovsky. *Nonlinear sloshing in tanks of complex geometrical shape*. Kiev: Naukova Dumka, 1975 (in Russian).
140. I.A. Lukovsky. *Variational method in the nonlinear problems of the dynamics of a limited liquid volume with free surface*. Oscillations of elastic constructions with liquid. Moscow: Volna, 260–264 (in Russian), 1976.
141. I.A. Lukovsky. *Variational method for solving the nonlinear problem on liquid sloshing in tanks of complicated shape*. Dynamics and stability of managed systems. Kiev: Institute of Mathematics, 1977 (in Russian).
142. I.A. Lukovsky. *Usage of variational principle to derivations of equations of the body-liquid dynamics*. Dynamics of spacecraft apparatus and study of the space. Moscow, Mashinostroenie, 182–194 (in Russian), 1986.
143. I.A. Lukovsky. *Introduction to nonlinear dynamics of rigid bodies with the cavities partially filled by a fluid*. Kiev: Naukova Dumka, 1990 (in Russian).
144. I.A. Lukovsky. On the theory of nonlinear sloshing of a weakly-viscous liquid. *Dopovidi of National Academy of Sciences of Ukraine*, (10):80–84 (in Ukrainian), 1997.
145. I.A. Lukovsky. On solving spectral problems on linear sloshing in conical tanks. *Dopovidi NANU*, (5):53–58 (in Ukrainian), 2002.
146. I.A. Lukovsky. A mathematical problem of wave liquid motions in reservoir with inclined walls. *Transactions of Institute of Mathematics of NASU*, 2(1):227–253 (in Ukrainian), 2005.

147. I.A. Lukovsky. *Nonlinear dynamics: mathematical models for rigid bodies with a liquid.* De Gruyter, 2015.
148. I.A. Lukovsky, M.Y. Barnyak, and A.N. Komarenko. *Approximate methods of solving the problems of the dynamics of a limited liquid volume.* Kiev: Naukova Dumka, 1984 (in Russian).
149. I.A. Lukovsky and A.N. Bilyk. *Forced sloshing in movable axisymmetric conical tanks.* Numerical-analytical studies of the dynamics and stability of multidimensional systems. Kiev: Institute of Mathematics, 12–26, (in Russian), 1985.
150. I.A. Lukovsky, O.V. Solodun, and A.N. Timokha. *Mathematical problems in nonlinear dynamics of conical reservoirs with liquid.* Naukova Dumka, 2019 (in Ukrainian).
151. I.A. Lukovsky and A.N. Timokha. *Variational method in nonlinear problems of the dymanics of a limited liquid volume.* Kiev: Institute of Mathematics of NASU, 1995 (in Russian).
152. I.O. Lukovsky and A.N. Bilyk. Study of forced nonlinear sloshing in conical tanks with a small apex angle. Applied problems in the dynamics and stability of mechanical systems. Kiev: Institute of Mathematics, 5–14, (in Russian) 1987.
153. I.O. Lukovsky and D.V. Ovchynnikov. Nonlinear mathematical model of the fifth order in the sloshing problem for a cylindrical tank. *Transactions of Institute of Mathematics of NASU.* 47:119–160, 2003.
154. I.O. Lukovsky and O.V. Solodun. A nonlinear model of liquid motions in cylindrical compartment tanks. *Dopovidi NANU*, (5):51–55 (in Ukrainian), 2001.
155. I.O. Lukovsky and O.V. Solodun. Study of the forced liquid sloshing in circular tanks by using a seven-mode model of the third order. *Transactions of Institute of Mathematics of NASU.* 47:161–179 (in Ukrainain), 2003.
156. O.D. Maggio and A. Rehm. Nonlinear free oscillations of a perfec fluid in a cylindrical container. *AIAA Symposium on Structural Dynamics and Aeroelasticity*, 30:156–161, 1965.
157. A. Martel, J.A. Nicolas, and J.M. Vega. Surface-wave damping in a brimful circular cylinder. *Journal of Fluid Mechanics*, 360:213–228, 1998.
158. E. Matta. Sistemi di attenuazione della risposta dinamica a massa oscillante solida e fluida: Ph.D. Thesis. Politecnico di Torino, Torino, (in Italian) 2002.
159. G. Mikishev and B. Rabinovich. *Dynamics of a Solid Body with Cavities Partially Filled with Liquid.* Mashinostroenie, 1968 (in Russian).
160. G. Mikishev and B. Rabinovich. *Dynamics of Thin-Walled Structures with Compartments Containing a Liquid.* Moscow: Mashinostroenie, 1971. (in Russian).
161. J. Miles. Nonlinear surface waves in closed basins. *Journal of Fluid Mechanics*, 75(3):419–448, 1976.
162. J. Miles. Internally resonant surface waves in circular cylinder. *Journal of Fluid Mechanics*, 149:1–14, 1984.
163. J. Miles. Resonantly forces surface waves in circular cylinder. *Journal of Fluid Mechanics*, 149:15–31, 1984.
164. J. Miles. Parametrically excited, progressive cross-waves. *Journal of Fluid Mechanics*, 186:129–146, 1988.

165. J. Miles. Parametrically excited, standing cross-waves. *Journal of Fluid Mechanics*, 186:119–127, 1988.
166. J. Miles. On Faraday waves. *Journal of Fluid Mechanics*, 248:671–683, 1993.
167. J. Miles. On Faraday waves. *Journal of Fluid Mechanics*, 269:372–372, 1994.
168. J.W. Miles. A note on interior vs. boundary-layer damping of surface waves in a circular cylinder. *Journal of Fluid Mechanics*, 364:319–323, 1998.
169. N. Moiseev. *Variational problems in the theory of oscillations of a liquid and a body with a liquid.* Moscow: Nauka, 1962 (in Russian).
170. N. Moiseev and A. Petrov. *Numerical methods for computing the eigenoscillations of a limited liquid volume.* Moscow: Computer Center of Academy of Sciences of USSR, 1966 (in Russian).
171. N.N. Moiseev. On the theory of nonlinear vibrations of a liquid of finite volume. *Journal of Applied Mathematics and Mechanics*, 22(5):860–872, 1958.
172. N. Moiseyev and V. Rumyantsev. Dynamic Stability of Bodies Containing Fluid. *Applied Physics and Engineering*, 6:251–306, 1968.
173. R. Moore and L. Perko. Inviscid fluid flow in an accelerating cylindrical container. *Journal of Fluid Mechanics*, 22:305–320, 1964.
174. J. Morand and R. Ohayon. *Fluid structure interaction. Applied numerical methods.* Chichester: John Wiley & Sons, 1995.
175. R. Multer. Exact nonliner model of wave generation. *Journal of the Hydraulics Division, ASME.*, 99 (HY1):31–46, 1973.
176. A. Myshkis, V. Babskii, and A. Kopachavskii. *Low-gravity fluid mechanics: Mathematical theory of capillary phenomena.* Berlin: Springer-Verlag, 1987.
177. T. Nakayama and K. Washizu. Nonlinear analysis of liquid motion in a container subjected to forced pitching oscillation. *International Journal for Numerical Methods in Engineering*, 15(8):1207–1220, 1980.
178. G. Narimanov. Motions of a solid body whose cavity is partl filled by a liquid. *Applied Mathematics and Mechanics (PMM)*, 20(1):21–38, 1956.
179. G. Narimanov. On the oscillations of liquid in the mobile cavities. *Izv. of the AS USSR, OTN.*, 10:71–74, (in Russian) 1957.
180. G.S. Narimanov, L.V. Dokuchaev, and I.A. Lukovsky. *Nonlinear dynamics of flying apparatus with liquid.* Moscow: Mashinostroenie, 1977 (in Russian).
181. H. Ockendon, J.H. Ockendon, and A.D. Johnson. Resonant sloshing in shallow water. *Journal of Fluid Mechanics*, 167:465–479, 1986.
182. J. Ockendon and H. Ockendon. Resonant surface waves. *Journal of Fluid Mechanics*, 59:397–413, 1973.
183. J.R. Ockendon, H. Ockendon, and D.D. Waterhouse. Multi-mode resonance in fluids. *Journal of Fluid Mechanics*, 315:317–344, 1996.
184. T. Okamoto and M. Kawahara. Two-dimensional sloshing analysis by Lagrangian finite element method. *Internatianal Journal for Numerical Methods in Fluids*, 11:453–477, 1990.
185. D. Okhotsimskii. On the theory of a body motions when there is a vacity partly filled by a liquid. *Applied Mathematics and Mechanics (PMM)*, 20(1):3–20, 1956.
186. M.A. Ostrogradsky. Mémoire sur la propagation des ondes dans un bassin cylindrique. *Mémoires a l'Academie Royale des Sciences, De l'Institut de France*, **III**:23–44, 1832.
187. L. Ovsyannikov, N. Makarenko, and V. Nalimov. *Nonlinear problems of the theory of surface and internal waves.* Novosibirsk: Nauka, 1985 (in Russian).

188. W. Penny and A. Price. Finite periodic stationary waves in a perfect liquid. *Philosophical Transactions of the Royal Society*, 244:254–284, 1952.
189. L. Perko. Large-amplitude motions of liquid-vapour interface in an accelerating container. *Journal of Fluid Mechanics*, 35:77–96, 1969.
190. A. Pilkevich. *Analysis of forced liquid sloshing in co-axial cylindrical reseivors.* Applied methods of studying the physical-mechanical processes. Kiev: Institute of Mathematics, 49–63 (in Russian), 1979.
191. A. Pilkevich. *Constructing the approximate solutions of nonlinear equations on wave motions of a liquid in a container.* Dynamics and stability of mechanical systems. Kiev: Institute of Mathematics, 16–21 (in Russian), 1980.
192. L. Prandtl. Erzeugung von Zirkulation beim Schütteln von Gefässen. *Zeitschrift für Angewandte Mathematik und Mechanik*, 29:8–9, 1949.
193. B. Rabinovich. Equations of perturbed motions of a solid body with a cylindrical cavity partly filled by a liquid. *Applied Mathematics and Mechanics (PMM)*, 20(1):39–49, 1956.
194. B. Rabinovich. *Introduction to dynamics of spacecraft.* Moscow: Mashinostroenie, 1975 (in Russian).
195. B. Ramaswamy, M. Kawara, and T. Nakayama. Lagrangian finite element method for the analysis of two-dimensional sloshing problems. *Internatianal Journal for Numerical Methods in Fluids*, 6(9):659–670, 1986.
196. I. Raynovskyy and A. Timokha. Resonant liquid sloshing in an upright circular tank performing a periodic motion. *Journal of Numerical and Applied Mathematics*, 122:71–82, 2016.
197. I. Raynovskyy and A. Timokha. Damped resonant steady-state sloshing in an upright circular tank. *Transactions of Institute of Mathematics of NAS of Ukraine*, 14(2):180–204, 2017.
198. I. Raynovskyy and A. Timokha. Damped steady-state resonant sloshing in a circular base container. *Fluid Dynamics Research*, 50 (Paper ID 045502.):1–27, 2018.
199. S. Rebouillat and D. Liksonov. Fluid-structure interaction in partially filled liquid cotainers: A comparative review of numerical approaches. *Computers & Fluids*, 39:739–746, 2010.
200. M. Reclari. *Hydrodynamics of orbital shaken bioreactors.* PhD Thesis, Ecole Polytechnique Federale de Lausanne, 2013.
201. M. Reclari, M. Dreyer, S. Tissot, D. Obreschkow, F.M. Wurm, and M. Farhat. Surface wave dynamics in orbital shaken cylindrical containers. *Physics of Fluids*, 26:1–11, 2014. Paper ID 052104.
202. O.F. Rognebakke and O.M. Faltinsen. Coupling of sloshing and ship motions. *Journal of Ship Research*, 47(3):208–221, 2001.
203. A. Royon-Lebeaud, E. Hopfinger, and A. Cartellier. Liquid sloshing and wave breaking in circular and square-base cylindrical containers. *Journal of Fluid Mechanics*, 577:467–494, 2007.
204. E. Seliger and G.B. Whitham. Variational principles in continuum mechanics. *Proceedings of the Royal Society*, 305:1–25, 1968.
205. P. Shankar and R. Kidambi. A modal method for finite amplitude, nonlinear sloshing. *Pramana*, 59(4):631–651, 2002.
206. Y.-L. Shao and O. Faltinsen. Fully-nonlinear wave-current-body interaction analysis by a harmonic polynomial cell (HPC) method. In *32nd International Conference on Ocean, Offshore and Arctic Engineering, Nantes, France, June 9-14, 2013*, 2013, Paper Number OMAE2013-10186.

207. Y.-L. Shao and O. Faltinsen. A harmonic polynomial cell (HPC) method for 3D laplace equation with application in marine hydromechanics. *Journal of Computational Physics*, 274:312–332, 2014.
208. A. Shvets and V. Sirenko. Peculiarities of transition to chaos in nonideal hydrodynamics systems. *Chaotic Modeling and Simulation*, 2:303–310, 2012.
209. J. Sicilian and J. Tegart. Comparison of FLOW-3D calculations with very large amplitude slosh data. In *Proc. ASME/JSME Pressure Vessels and Piping Conference, Honolulu, HI, July 23-27, 1989 (A90-32339 13-31)*. New York, American Society of Mechanical Engineers, 23–30, 1989.
210. F. Solaas. *Analytical and numerical studies of sloshing in tanks: Ph.D Thesis*. PhD Thesis, Norwesian Institute of Technology, 1995.
211. F. Solaas and O. Faltinsen. Combined numerical and analytical solution for sloshing in two-dimensional tanks of general shape. *Journal of Ship Research*, 41(2):118–129, 1997.
212. A. V. Solodun and A. N. Timokha. *The Narimanov–Moiseev multimodal analysis of nonlinear sloshing in circular conical tanks*, volume 177 of *Studies in Systems, Decision and Control: Applied Mathematical Analysis: Theory, Methods, and Applications*, 267–309. Springer International Publishing, 2020.
213. A. Stofan and A. Armsted. *Analytical and experimental investigation of forces and frequencies resulting from liquid sloshing in a spherical tank: Tech. Rep. D-1281: NASA*. Lewis Research Center, Cleveland, Ohio, 1962.
214. V. Stolbetsov. Nonsmall liquid oscillations in a right circular cylinder. *Fluid Dynamics*, 2(2):41–45, 1967.
215. V. Stolbetsov. Oscillations of liquid in a vessel in the form of a rectangular parallelepiped. *Fluid Dynamics*, 2(1):44–49, 1967.
216. V. Stolbetsov. Equations of nonlinear oscillations of a container partially filled with a liquid. *Fluid Dynamics*, 4(2):95–99, 1969.
217. V. Stolbetsov and V. Fishkis. Equations of nonlinear oscillations of a container partially filled with a liquid. *Fluid Dynamics*, 3(5):79–81, 1968.
218. T. Su and Y. Wang. *Numerical simulation of three-dimensional large amplitude liquid sloshing in cylindrical tanks subjected to arbitrary excitations*. Proc., 1990 Pressure Vessels and Piping Conf., ASME, New York, N.Y., 127–148, 1990.
219. I.E. Sumner. *Experimental investigations of stability boundaries for planar and nonplanar sloshing in spherical tanks: Tech. rep.: NASA, TN D-3210*. Lewis Research Center, Cleveland, Ohio, 1966.
220. I.E. Sumner and A.J. Stofan. *An experimental investigation of the viscous damping of liquid sloshing in spherical tanks: Tech. rep.: NASA, TN D-1991*. Lewis Research Center, Cleveland, Ohio, 1963.
221. P. Sun, F. Ming, and A Zhang. Numerical simulation of interactions between free surface and rigid body using a robust SPH method. *Ocean Engineering*, 98:23–49, 2015.
222. H. Takahara, K. Hara, and T. Ishida. Nonlinear liquid oscillation in a cylindrical tank with an eccentric core barrel. *Journal of Fluids and Structures*, 35:120–132, 2012.
223. H. Takahara and K. Kimura. Frequency response of sloshing in an annular cylindrical tank subjected to pitching excitation. *Journal of Sound and Vibration*, 331(13):3199–3212, 2012.

224. A. Timokha and I. Raynovskyy. The damped sloshing in an upright circular tank due to an orbital forcing. *Reports of the National Academy of Sciences of Ukraine*, (2):48–53, 2017.
225. A.N. Timokha and I.A. Raynovskyy. Resonant steady-state sloshing in upright tanks: effect of three-dimensional excitations and viscosity. In *OMAE – 37th International Conference on Ocean, Offshore & Arctic Engineering. June 17-22, 2018*, 2018, Paper Number: OMAE2018-77534.
226. V.A. Trotsenko. *Oscillations of a liquid in mobile containers with ribs*. Kiev: Institute of Mathematics, 2006 (in Russian).
227. M. Utsumi. Theoretical determination of modal damping ratio of sloshing using a variational method. *Journal of Pressure Vessel Technology*, 133(Art. No. 011301-1):1–10, 2011.
228. I.N. Vekua. On completeness of a system of harmonic polynomials in space. *Doklady Akademii Nauk SSSR*, 90:495–498, 1953 (in Russian).
229. I.N. Vekua. *New methods for solving elliptic equations*. New York: Interscience Publishers John Wiley & Sons, Inc., 1967.
230. J. H. G. Verhagen and L. Wijngaarden. Non-linear oscillations of fluid in a container. *Journal of Fluid Mechanics*, 22(4):737–752, 1965.
231. L. Wang, Z. Wang, and Y. Li. A SPH simulation on large-amplitude sloshing for fluids in a two-dimensional tank. *Earthquake Engineering and Engineering Vibration*, 12(1):135–142, 2013.
232. D.D. Waterhouse. Resonant sloshing near a critical depth. *Journal of Fluid Mechanics*, 21:313–318, 1994.
233. W. Weheliye, M. Yianneskis, and A. Ducci. On the fluid dynamics of shaken bioreactors – flow characterization and transition. *AIChE Journal*, 59:334–344, 2013.
234. G.B. Whitham. Non-linear dispersion of water waves. *Journal of Fluid Mechanics*, 27:399–412, 1967.
235. G.B. Whitham. *Linear and nonlinear waves*. New York: Interscience, 1974.
236. N.M. Wigley. Asymptotic expansions at a corner of solutions of mixed boundary value problems (Asymptotic expansions at corner of solutions of elliptic second-order partial differential equations in two variables). *Journal of Mathematics and Mechanics*, 13:549–576, 1964.
237. N.M. Wigley. Mixed boundary value problems in plane domains with corners. *Mathematische Zeitschrift*, 115:33–52, 1970.
238. G.X. Wu. Second-order resonance of sloshing in a tank. *Ocean Engineering*, 34:2345–2349, 2007.
239. Y. Yan-Sheng, M. Xing-Rui, and W. Ben-Li. Multidimensional modal analysis of liquid Nonlinear sloshing in right circular cylindrical tank. *Applied Mathematics and Mechanics*, 28(8):1997–2018, 2007.
240. C. Zhang. Application of an improved semi-Lagrangian procedure to fully-nonlinear sloshing in right circular cylindrical tank. *Applied Ocean Research*, 51:74–92, 2015.
241. C. Zhang, Y. Li, and Q. Meng. Fully nonlinear analysis of second-order sloshing resonance in a three-dimensional tank. *Computers & Fluids*, 116:88–104, 2015.

Index

acceleration
    centripental, 29
    convective, 30
    Coriolis, 29
    gravity, 13
action, 26
adaptive
    approximation of the hydrodynamic moment, 83
    modal equations, 8, 60, 66
    multimodal theory, 77
angular momentum, 29
asymptotic periodic solutions, 89
asymptotic relations
    Narimanov-Moiseev, 70
asymptotically equivalent elliptic tank motion, 92
atmospheric pressure, 15
average
    energy loss of a standing wave, 50
    total energy, 51

Bateman-Luke
    action, 17
    formalism, 5
    formulation, 21
    principle, 17
Bernoulli equation, 15
    inertial coordinate system, 14
    non-inertial coordinate system, 15
Bernoulli lemniscate, 102
Bessel function of the first kind, 57
Bond number, 13, 52, 110
boundary conditions
    dynamic, 15
    kinematic, 16

boundary layer
    damping, 52
    thickness, 13, 52
Boussinesq, 3
Boussinesq models, 6, 8
breaking waves, 12, 109, 122
Bubnov-Galerkin method, 89
bulk
    damping, 50
    viscosity, 8
    viscosity coefficient, 50

capillary meniscus, 13
Cartano's theorem, 114
Cauchy conditions, 17, 43
classification of steady-state waves, 9
    for longitudinal forcing, 99
    for orbital forcing, 111
Clebsch potentials, 19
commensurate spectrum, 8
compartment tank, 10
completely filled tank, 29, 31, 46
conservation of mass, 16
continuity equation, 14, 20
coordinate system
    Cartesian, 14
    cylindrical, 27
    Earth-fixed, 13
    tank-fixed, 13
criterion [standing, clockwise/counterclockwise swirling], 36, 97
critical depth, 7, 8, 76
cylindrical coordinate system, 37

damped harmonic oscillator, 48
damping
    bulk viscosity, 51

coefficient, 103
coefficients, 49
damping ratio, 10
laminar viscous layer, 50
ratio, 49, 50
surface tension, 52
viscous effects, 43
dissipative phenomena, 50
divergence theorems, 20
dominant higher harmonics, 8
dot-differentiation, 57
dynamic
contact angle, 12
modal subsystem, 59
dynamic boundary conditions, 15

eigenfrequencies, 57
elliptic forcing, 112
energy, 50
energy dissipation
bulk damping, 52
viscous boundary-layer, 51
energy loss
averaged total enegy, 51
standing wave, 50
equivalent elliptic forcing, 111
Euler formula, 15
Euler-Lagrange equation, 1, 24

Faraday waves, 5
free-standing wave, 34
free-surface
dynamic condition, 40
elevation, 41
kinematic condition, 40
problem, 13
frequency parameter, 96

Galileo number, 50
Gauss' theorem, 18
gravity potential, 13
Green's formula, 18

holonomic constraint, 16
hybrid mechanical system, 1

hydrodynamic
coefficients, 42, 47, 72, 123
force, 76
generalised coordinates, 1, 4, 23, 70
generalised velocities, 70
moment, 76
momentum, 28
hydrodynamic force, 48
adaptive approximation, 76
hydrodynamic instability, 103
hydrodynamic moment, 48
adaptive approximation, 83
Narimanov-Moiseev approximation, 85
hydrostatic liquid mass centre, 29

incompressible liquid, 13, 14
inertia tensor, 25, 45
instantaneous angular velocity, 15
integral amplitudes, 96
irregular sloshing, 99, 100
irrotational flows, 17

Keulegan method, 50
kinematic
boundary conditions, 16, 20
modal subsystem, 59
Kordeweg-de-Vries, 3
Kronecker delta, 57

Lagrange multiplier, 19
Lagrangian, 23, 24
laminar viscous layer, 50
Landau-Lifshitz formula, 52
Laplace equation, 14
linear
natural sloshing modes and frequencies, 33
linearised sloshing problem, 40
liquid
mass, 48
mass centre, 29

liquid density, 13
liquid mass centre, 29, 76
logarithmic decrement, 49
long-time scale, 44
longitudinal forcing, 99, 100
lowest-order approximation, 89, 97
Lukovsky formula, 3, 7
    force, 44
    hydrodynamic force, 28, 29
    hydrodynamic force linearised, 44
    hydrodynamic moment, 31
    moment, 28, 45
Lyapunov method, 10, 91

mass conservation, 16
mean free surface, 13
mean square approximation, 108
Miles equations, 5
modal equations, 47
    adaptive, 60, 66
    linear, 40, 42
    Miles-Lukovsky, 27, 59, 60
    Narimanov-Moiseev, 70, 92, 111
Moiseev
    asymptotics, 70
    detuning, 4, 70, 92
multimodal method, 1, 2, 40, 57
    linear, 2, 39, 40
multimodal theory
    linear, 46

Narimanov technique, 5
Narimanov-Moiseev
    approximation, 85
    asymptotic relations, 70, 76
    asymptotics, 7, 10
natural sloshing
    frequencies, 3, 34, 37
    frequency, 70
    modes, 3, 34, 37, 57
    periods, 34
    theory, 3

Neumann boundary problem, 24
nodal line, 38
non-inertial coordinate system, 14, 28
    time derivative, 28
nonconformal mapping technique, 6, 10
nondimensional gravity acceleration, 55, 56
nondimensional natural sloshing modes, 57

orbital tank motion, 88
orthogonality condition, 34
outer normal to the free surface, 16

pendulum analogy, 36
perforated screen, 8
periodic (steady-state) solutions, 60
periodicity condition, 38
Perko method, 5, 6
phase lags, 96
phase-lag difference, 112
pitch, 55
planar wave, 96, 100, 101
potential flows, 30
pressure, 4, 14, 16

Q-factor, 49, 50

radial wave mumbers, 57
rectangular tank, 7, 9, 35
response curves
    phase-lag, 104
    wave-amplitude, 99, 103
Reynolds
    hypothesis, 7
    transport theorem, 18, 20
roll, 55
roof impact, 8
rotational flows, 19

second-order approximation, 80
secondary resonance, 4, 7, 74, 92

secular
    equations, 89
        alternative, 95, 97
    system, 90
separation
    slow and fast time, 91
    spatial variables, 35, 37
singular behaviour at sharp edges, 4
slow time scale, 91
solidity ratio, 9
solvability condition, 90
spectral
    boundary problem, 33
    matrix problem, 91
    parameter, 57
square base container, 36
squares-like wave, 36
stability, 91, 92
    condition, 91
standing wave, 35
star-differentiation, 57
steady-state
    sloshing, 102
    solution, 42
    wave, 42
    wave modes
        damped sloshing, 100
        undamped sloshing, 99
Stokes mode
    combined, 36
    standing wave, 36
Stokes-Joukowski potentials, 3, 23, 24, 41, 47, 57
    approximate generalised, 78
    generalised, 24
surface tension, 13, 52
surge, 55
sway, 55
swirling, 36, 37, 100, 101
    clockwise, 100, 112
    clockwise/counterclockwise, 36, 97
    co-directed, 116
    counter-directed, 116, 117
    counterclockwise, 100, 112
system of ordinary differential equations, 60

tap water, 53
total averaged energy, 49
transient waves, 42
translational velocity, 14

upright annular cylindrical tank, 10
upright circular tank, 36

variational principle
    Lagrangian, 2
velocity
    field, 14, 19
    potential, 14, 56
    potential (the Fourier expansion), 41
viscous boundary layer, 8, 107
    laminar, 50
volume conservation condition, 34, 41

Printed in the United States
By Bookmasters